INTRODUCTION TO
GRAPH THEORY
With Solutions to Selected Problems

Other World Scientific Titles by the Author

Principles and Techniques in Combinatorics
ISBN: 978-981-02-1114-1
ISBN: 978-981-02-1139-4 (pbk)

Chromatic Polynomials and Chromaticity of Graphs
ISBN: 978-981-256-317-0
ISBN: 978-981-256-383-5 (pbk)

Counting
ISBN: 978-981-238-063-0
ISBN: 978-981-238-064-7 (pbk)

Counting: Supplementary Notes and Solutions Manual
ISBN: 978-981-256-915-8 (pbk)

Counting
Second Edition
ISBN: 978-981-4401-90-6
ISBN: 978-981-4401-91-3 (pbk)

Counting: Solutions Manual
Second Edition
ISBN: 978-981-4401-94-4 (pbk)

Introduction to Graph Theory: H3 Mathematics
ISBN: 978-981-270-525-9
ISBN: 978-981-270-386-6 (pbk)

Introduction to Graph Theory: Solutions Manual
ISBN: 978-981-277-175-9 (pbk)

Graph Theory: Undergraduate Mathematics
ISBN: 978-981-4641-58-6
ISBN: 978-981-4641-59-3 (pbk)

INTRODUCTION TO
GRAPH THEORY
With Solutions to Selected Problems

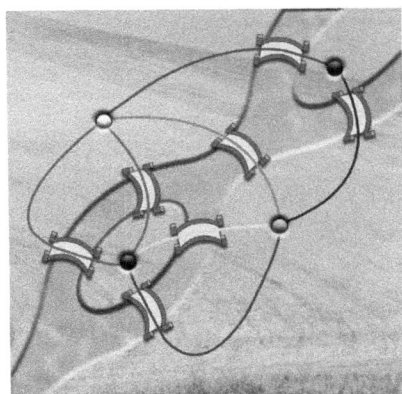

Khee Meng Koh
National University of Singapore, Singapore (Emeritus)

Fengming Dong
Nanyang Technological University, Singapore

Eng Guan Tay
Nanyang Technological University, Singapore

World Scientific

NEW JERSEY · LONDON · SINGAPORE · BEIJING · SHANGHAI · HONG KONG · TAIPEI · CHENNAI · TOKYO

Published by

World Scientific Publishing Co. Pte. Ltd.

5 Toh Tuck Link, Singapore 596224

USA office: 27 Warren Street, Suite 401-402, Hackensack, NJ 07601

UK office: 57 Shelton Street, Covent Garden, London WC2H 9HE

Library of Congress Control Number: 2023053687

British Library Cataloguing-in-Publication Data
A catalogue record for this book is available from the British Library.

INTRODUCTION TO GRAPH THEORY
With Solutions to Selected Problems

ISBN 978-981-12-8481-6 (hardcover)
ISBN 978-981-12-8501-1 (paperback)
ISBN 978-981-12-8482-3 (ebook for institutions)
ISBN 978-981-12-8483-0 (ebook for individuals)

For any available supplementary material, please visit
https://www.worldscientific.com/worldscibooks/10.1142/13637#t=suppl

Desk Editor: Tan Rok Ting

Printed in Singapore

Preface

Discrete Mathematics is a branch of mathematics dealing with finite or countable processes and elements. Graph theory is an area of Discrete Mathematics which studies configurations (called graphs) consisting of a set of nodes (called vertices) interconnecting by lines (called edges). From humble beginnings and almost recreational type problems, Graph Theory has found its calling in the modern world of complex systems and especially of the computer. Graph Theory and its applications can be found not only in other branches of mathematics, but also in scientific disciplines such as engineering, computer science, operational research, management sciences and the life sciences. Since computers require discrete formulation of problems, Graph Theory has become an essential and powerful tool for engineers and applied scientists, in particular, in the area of designing and analyzing algorithms for various problems which range from designing the itineraries for a shipping company to sequencing the human genome in the life sciences.

Graph Theory shows its versatility in the most surprising areas. Recently, the connectivity of the World Wide Web and the number of links needed to move from one webpage to another has been remarkably modeled with graphs, thus opening the real world internet connectivity to more rigorous studies. These studies form part of research into the phenomena of the property of a 'small world' even in huge systems such as the aforementioned internet and global human relationships (in the so-called 'Six Degrees of Separation').

This book is intended as a general introduction to Graph Theory. The first edition was written as a resource book for junior college students and teachers reading and teaching the subject at the Cambridge H3 Advanced Level in the Singapore Mathematics curriculum for Junior College. The topic coverage however is also suitable for a beginning course in undergraduate mathematics. The second edition now includes selected solutions and hints for the book exercises, thus making it a suitably complete first undergraduate Graph Theory textbook and reference. In addition, the variety of problems and applications in the book are not only useful for building up an aptitude in Graph Theory but are a rich source for honing basic skills and techniques in general problem solving and logical thinking.

Certain features of this book are worth mentioning. The book is written with great care that concepts are explained clearly and developed properly; it strives to be student friendly, and at the same time be mathematically rigorous. At suitable junctures, questions are inserted for discussion. This is to ensure that the reader understands the preceding section fully before proceeding on to new ideas and concepts. There are many questions in the Exercise component following most sections. Some are exercises intended for reinforcing what is learnt earlier while others test the full range of understanding and problem solving in the concepts acquired. Proofs of most important theorems are given in their full mathematical rigour. Each chapter concludes with applications of the concepts in real-life, which are added for general interest and as substantiation of the usefulness of Graph Theory concepts. References are cited in full at the end of the book in the References section and they are indexed with the first letter of the first author's name within square parentheses. For example, [E] is for a paper by Euler. The symbol □ is used to indicate the end of a proof or a result stated without proof. Challenging problems are indicated with the symbol (+).

Chapter 1 covers the fundamental concepts and basic results in Graph Theory tracing its history from Euler's solution of the problem of the Seven Bridges of Königsberg. Fundamental concepts include those of graphs, multigraphs, vertex degrees, paths, cycles and connectedness.

When are two graphs the 'same'? Following the style of Chapter 1, Chapter 2 further exposes the student to the rigour of mathematics in constructing a theory through definitions and theorems. Since two graphs may look different and yet 'function' similarly, the empirical perspective that mathematics students are so accustomed to needs to be reconsidered. Thus, congruence is defined in terms of isomorphism rather than a vague notion of shape thus enabling a 'handle' to compare graphs. This rigour and mathematical method of definitions and theorems continues throughout the whole book.

In Chapter 3, we introduce two important families of graphs, namely trees and bipartite graphs. A tree, in some sense, forms the 'skeleton' of a connected graph and in general, a forest of trees forms the 'skeleton' of any graph. Thus, the structure and properties of trees are very important. Bipartite graphs are another family of graphs that have found applications in many real-life situations such as matching a group of job seekers with a set of potential jobs under certain conditions.

Are four colours sufficient to colour any map? This question had frustrated many great mathematicians for over a century. Chapter 4 introduces the concept of vertex colouring which rephrases the question more simply. The notion of chromatic number (minimum number of colours used) is presented and an algorithm and some techniques to estimate or enumerate it are discussed. Interesting applications of vertex-colouring to scheduling problems are given in some detail.

Chapter 5 expands on the concept of matchings in bipartite graphs introduced in Chapter 3. Here we have a beautiful classical result in Graph Theory — Hall's Theorem. The necessity of the condition is trivial but the insight leading to the condition and the proof of its sufficiency exhibits the creativity of good mathematics. Hall's Theorem is used to determine the existence of a complete matching and this is used to good effect in the Marriage Problem and to find a system of distinct representatives (SDR).

Chapter 6 returns the reader to Euler's seminal work on the Bridges of Königsberg. Euler is memorialized for his contribution by having graphs with the property that one can have a walk that traverses all edges exactly once and that returns to the starting vertex named after him - Eulerian multigraphs. This chapter gives a fuller treatment of Eulerian multigraphs. It also discusses an apparently similar concept, that of graphs with the property that one can have a walk that visits all vertices exactly once and that returns to the starting vertex. This kind of graphs is called Hamiltonian, named after another mathematical giant, William Rowan Hamilton.

The last chapter is a necessary addition in an introductory book on Graph Theory. Chapter 7 studies graphs with 'directions' indicated on the edges. Such are called directed graphs or digraphs. This addition to Graph Theory suitably models many situations where relationships between items (vertices) are directional. The chapter covers some basic concepts and provides some detail on the most basic of digraphs which are tournaments.

We would like to thank Dr. Kho Tek Hong, Dr. K. L. Teo, Ms Goh Chee Ying and Mr Soh Chin Ann for reading through the draft of the first edition and checking through the problems.

For those who find this introductory book interesting and would like to know more about the subject, a recommended list of publications for further reading is provided at the end of this book.

Koh Khee Meng
Dong Fengming
Tay Eng Guan
July 2023

Notation

$\mathbb{N} = \{1, 2, 3, \cdots\}$

$|S| =$ the number of elements in the finite set S

$\binom{n}{r} =$ the number of r-element subsets of an n-element set $= \frac{n!}{r!(n-r)!}$

$B \setminus A = \{x \in B | x \notin A\}$, where A and B are sets

$\bigcup_{i \in I} S_i = \{x | x \in S_i \text{ for some } i \in I\}$, where S_i is a set for each $i \in I$

In what follows, G and H are multigraphs, and D is a digraph.

$V(G) :$ the vertex set of G

$E(G) :$ the edge set of G

$v(G) :$ the number of vertices in G or the order of G

$e(G) :$ the number of edges in G or the size of G

$V(D) :$ the vertex set of D

$E(D) :$ the arc set of D

$v(D) :$ the number of vertices in D or the order of D

$e(D) :$ the number of arcs in D

$x \to y :$ x is adjacent to y, where x, y are vertices in D

$x \nrightarrow y :$ x is not adjacent to y, where x, y are vertices in D

$G \cong H :$ G is isomorphic to H

$A(G) :$ the adjacency matrix of G

$\overline{G} :$ the complement of G

$[A] :$ the subgraph of G induced by A, where $A \subseteq V(G)$

$e(A, B) :$ the number of edges in G having an end in A and the other in B, where $A, B \subseteq V(G)$

$G - v :$ the subgraph of G obtained by removing v and all edges incident with v from G, where $v \in V(G)$

$G - e$: the subgraph of G obtained by removing e from G, where $e \in E(G)$

$G - F$: the subgraph of G obtained by removing all edges in F from G, where $F \subseteq E(G)$

$G - A$: the subgraph of G obtained by removing each vertex in A together with the edges incident with vertices in A from G, where $A \subseteq V(G)$

$G + xy$: the graph obtained by adding a new edge xy to G, where $x, y \in V(G)$ and $xy \notin E(G)$

$N(u) = N_G(u)$: the set of vertices v such that $uv \in E(G)$

$N(S) = \bigcup_{u \in S} N(u)$, where $S \subseteq V(G)$

$d(v) = d_G(v)$: the degree of v in G, where $v \in V(G)$

$id(v)$: the indegree of v in D, where $v \in V(D)$

$od(v)$: the outdegree of v in D, where $v \in V(D)$

$d(u, v)$: the distance between u and v in G, where $u, v \in V(G)$

$c(G)$: the number of components in G

$\delta(G)$: the minimum degree of G

$\Delta(G)$: the maximum degree of G

$\chi(G)$: the chromatic number of G

$\alpha(G)$: the independence number of G

$G + H$: the join of G and H

$G \cup H$: the disjoint union of G and H

kG : the disjoint union of k copies of G

$G(D)$: the underlying graph of D

$n_G(H)$: the number of subgraphs in G which are isomorphic to H

C_n : the cycle of order n

K_n : the complete graph of order n

N_n : the null graph or empty graph of order n

P_n : the path of order n

W_n : the wheel of order n, $W_n = C_{n-1} + K_1$

$K(p, q)$: the complete bipartite graph with a bipartition (X, Y) such that $|X| = p$ and $|Y| = q$

Contents

Chapter 1

Fundamental Concepts and Basic Results

1.1   The Königsberg bridge problem

In an old city of Eastern Prussia, named *Königsberg*, there was a river, called River Pregel, flowing through its centre. In the 18th century, there were seven bridges over the river connecting the two islands (B and D) and two opposite banks (A and C) as shown in Figure 1.1.

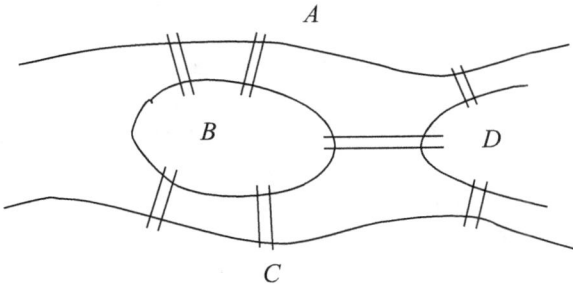

Figure 1.1

It was said that the people in the city had always amused themselves with the following problem:

Starting with any one of the four places A, B, C or D as shown in Figure 1.1, is it possible to have a walk which passes through each of the seven bridges once and only once, and return to where you started?

No one could find such a walk; and after a number of tries, people believed that it was simply not possible, but no one could prove it either.

Leonhard Euler, the greatest mathematician that Switzerland has ever produced, was told of the problem. He noticed that the problem was very much different in nature from the problems in traditional geometry, and instead of considering the original problem, he studied its much more general version which encompassed any number of islands or banks, and any number of bridges connecting them. His finding was contained in the article [E] (the English translation of its title is: *The solution of a problem to the geometry of position*) published in 1736. As a direct consequence of his finding, he deduced the impossibility of having such a walk in the Königsberg bridge problem. This was historically the first time a proof was given from the mathematical point of view.

How did Euler generalize the Königsberg bridge problem? How did he solve his more general problem? What was his finding?

1.2   Multigraphs and graphs

Euler observed that the Königsberg bridge problem had nothing to do with traditional geometry where the measurements of lengths and angles, and relative locations of vertices count. How large the islands and banks are, how long the bridges are, and whether an island is at the south or north of a bank are immaterial. The key ingredients are whether the islands or banks are connected by a bridge, and by how many bridges.

Euler's idea was essentially as follows: represent the islands or banks by '**dots**', one for each island or bank, and two dots are joined by k '**lines**' (not necessarily straight), where $k \geq 0$, when and only when the respective islands or banks represented by the dots are connected by k bridges. Thus the situation for the Königsberg bridge problem is represented by the diagram in Figure 1.2.

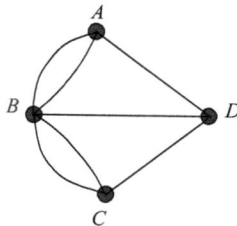

Figure 1.2

The diagram in Figure 1.2 is now known as a **multigraph**. *Intuitively,* a **multigraph** is a diagram consisting of 'dots' and 'lines', where each line joins some pair of dots, and two dots may be joined by no lines or any number of lines. More formally, we call a 'dot' a **vertex** (plural, vertices) and call a 'line' an **edge**.

For instance, in the multigraph of Figure 1.2, there are four vertices and seven edges, where each edge joins some pair of vertices; vertices A and C are not joined by any edges, A and D are joined by one edge, and B and C are joined by two edges, etc.

Note that the sizes and the relative locations of dots (vertices), and the lengths of the lines (edges) are immaterial. Only the 'linking relations' among the vertices and the number of edges that join two vertices count. Thus, the situation for the Königsberg bridge problem can equally well be represented by the multigraph of Figure 1.3.

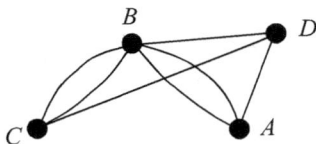

Figure 1.3

Let us give more examples of multigraph which represent certain situations in different nature.

Example 1.2.1. *There were six people: A, B, C, D, E and F in a party and several handshakes among them took place. Suppose that*

> *A shook hands with B, C, D, E and F,*
> *B, in addition, shook hands with C and F,*
> *C, in addition, shook hands with D and E,*
> *D, in addition, shook hand with E,*
> *E, in addition, shook hand with F.*

This situation can be clearly shown by the multigraph in Figure 1.4, where people are represented by vertices and two vertices are joined by an edge whenever the corresponding persons shook hands.

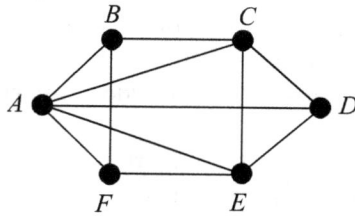

Figure 1.4

Example 1.2.2.   *The diagram in Figure 1.5 is a multigraph which shows the availability of flights operated by an airline company between a number of cities. The vertices represent the cities, and two vertices are joined by an edge if there is a flight available between the two corresponding cities.*

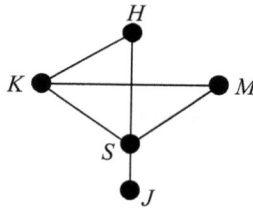

Figure 1.5

Example 1.2.3.   *The diagram in Figure 1.6 is a multigraph which models a job-application situation. The vertices are divided into two parts: X and Y, where the vertices in X represent the applicants, while those in Y represent the jobs available. A vertex in X is joined to a vertex in Y by an edge if the corresponding applicant applies for the corresponding job.*

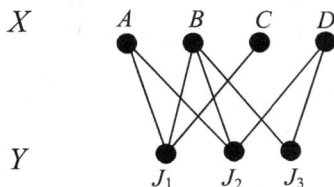

Figure 1.6

Question 1.2.1. *Give three examples from our everyday life where the situations can be modeled by multigraphs.*

It is noted that in the three multigraphs shown in Figures 1.4 to 1.6, every two vertices are joined by at most one edge (that is, either no edges or exactly one edge). These situations are different from the multigraph in Figure 1.2 (or Figure 1.3) where there are vertices joined by more than one edge. To distinguish them, we call the diagrams in Figures 1.4 to 1.6 **simple graphs,** or simply, **graphs.** Thus the diagram in Figure 1.2 (or Figure 1.3) is a multigraph, but not a (simple) graph.

Let us consider another example.

Example 1.2.4. *In the diagram shown in Figure 1.7, there are*

- *four vertices: u, v, w and z, and*
- *eight edges: f_1 and f_2 joining u and v; e_1, e_2 and e_3 joining w and z; h_1 joining v and w; h_2 joining u and w; h_3 joining v to itself.*

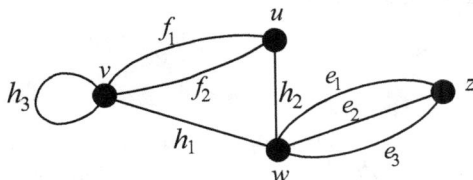

Figure 1.7

Two or more edges joining the same pair of vertices are called **parallel edges**. Thus, in Figure 1.7, f_1 and f_2 are parallel edges; e_1, e_2 and e_3 are parallel edges.

Any edge joining a vertex to itself is called a **loop**. Thus, in Figure 1.7, h_3 is a loop.

Remarks. (1) In this book, we shall not consider 'loops' in any diagram of vertices and edges unless otherwise stated. A diagram with the existence of parallel edges is **not** a **(simple) graph**. Another example of a multigraph which is **not** a (simple) graph is shown in Figure 1.8.

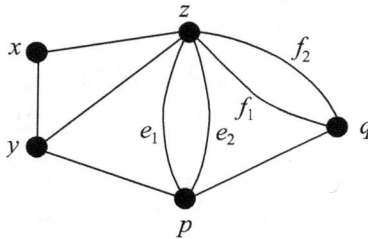

Figure 1.8

(2) Bear in mind that a 'graph' or a 'multigraph' in Graph Theory is not a geometrical figure. Thus we do not consider

- the size of a 'dot',
- the location of a vertex, and
- the shape of an edge.

(3) When there is only one edge joining a pair of vertices, say a and b, we may denote this edge by ab. For example, the edge h_1 in Figure 1.7 can also be denoted by vw.

We now give formal definitions of 'graph' and 'multigraph'.

A **multigraph** G consists of a non-empty finite set $V(G)$ of vertices together with a finite set $E(G)$ (possibly empty) of edges such that

(1) each edge joins two distinct vertices in $V(G)$ and

(2) any two distinct vertices in $V(G)$ are joined by a finite number (including zero) of edges.

The sets $V(G)$ and $E(G)$ are called the **vertex set** and the **edge set** of G respectively.

The number of vertices in G, denoted by $v(G)$, is called the **order** of G (thus $v(G) = |V(G)|$). The number of edges in G, denoted by $e(G)$, is called the **size** of G (thus $e(G) = |E(G)|$).

A multigraph G is called a (simple) **graph** if any two vertices in $V(G)$ are joined by at most one edge (that is, either they are not joined by an edge or joined by exactly one edge).

(a) It follows from the above definitions that

(i) every graph is a multigraph but not vice versa and

(ii) no loops are allowed in any multigraph.

When a concept is defined or a statement is made for multigraphs, they are also valid, in particular, for graphs.

(b) If e is the only edge joining two vertices u and v, then we may write $e = uv$ or $e = vu$. The ordering of u and v in the expression is immaterial.

Example 1.2.5.   *Let G be the multigraph shown in Figure 1.8. Then*
$V(G) = \{x, y, z, p, q\}$,
$E(G) = \{xy, xz, yz, yp, e_1, e_2, f_1, f_2, pq\}$,
$v(G) = 5$ *and* $e(G) = 9$.
Let H be the graph shown in Figure 1.4. Then
$V(H) = \{A, B, C, D, E, F\}$,
$E(H) = \{AB, AC, AD, AE, AF, BC, BF, CD, CE, DE, EF\}$,
$v(H) = 6$ *and* $e(H) = 11$.

Question 1.2.2. *Let G be the multigraph shown below. Find $V(G), E(G), v(G)$ and $e(G)$.*

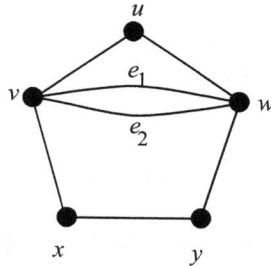

Question 1.2.3. *Let H be the graph with $V(H) = \{a, b, c, x, y, z\}$ and $E(H) = \{ab, ay, bx, by, cx, cz, xz, yz\}$. Find $v(H)$ and $e(H)$, and draw a diagram of H.*

Matrices and multigraphs

As discussed earlier, a multigraph G can be represented by a diagram consisting of 'dots' and 'lines', and can be defined in terms of its vertex set $V(G)$ and edge set $E(G)$. Multigraphs can also be represented by matrices in various ways. In what follows, we introduce one of them.

Example 1.2.6. *Let G be the multigraph shown below, where its four vertices are named as v_1, v_2, v_3 and v_4.*

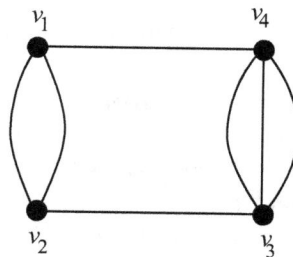

Consider also the following 4×4 *matrix* A:

$$A = \begin{pmatrix} 0 & 2 & 0 & 1 \\ 2 & 0 & 1 & 0 \\ 0 & 1 & 0 & 3 \\ 1 & 0 & 3 & 0 \end{pmatrix}$$

Can you find any relation between G *and* A?

What is the value of the $(1,2)$*-entry in* A? *It is* '2'. *How many edges in* G *join* v_1 *and* v_2? *There are* '2' *also.*

How many edges in G *join* v_3 *and* v_4? *There are* '3'. *What is the value of the* $(3,4)$*-entry in* A? *It is* '3' *also.*

Indeed, it is observed that the value of the (i,j)-entry in A is the number of edges in G joining v_i and v_j, where $i,j \in \{1,2,3,4\}$. Note that the value of each (i,i)-entry (that is, a diagonal entry) in A is '0' as there is no edge in G joining v_i to itself. We call A the **adjacency matrix** of G. Two vertices are adjacent if they are joined by an edge. Evidently, the matrix is dependent on the labelling of the vertices.

Let G be a multigraph of order n with $V(G) = \{v_1, v_2, \cdots, v_n\}$. The **adjacency matrix** of G is the $n \times n$ matrix

$$A(G) = (a_{i,j})_{n \times n},$$

where $a_{i,j}$, the (i,j)-entry in $A(G)$, is the number of edges joining v_i and v_j for all $i,j \in \{1,2,\cdots,n\}$.

Question 1.2.4.   *Let G be the multigraph shown below.*

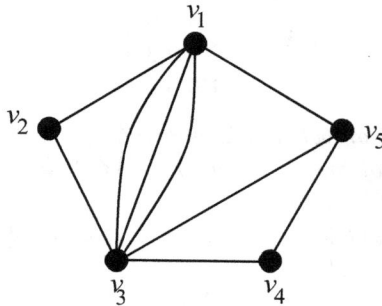

(i) *Find $A(G)$.*

(ii) *Is $A(G)$ symmetric (i.e., (i,j)-entry $= (j,i)$-entry)?*

(iii) *What is the sum of the values of the entries in each row (respectively, column)?*

(iv) *What is your interpretation of the 'sum' obtained in (iii)?*

Question 1.2.5.   *The adjacency matrix of a multigraph G is given below:*

$$A = \begin{pmatrix} 0 & 2 & 1 & 0 & 1 \\ 2 & 0 & 1 & 0 & 0 \\ 1 & 1 & 0 & 3 & 2 \\ 0 & 0 & 3 & 0 & 0 \\ 1 & 0 & 2 & 0 & 0 \end{pmatrix}$$

Draw a diagram of G.

Remark. There are many ways of storing multigraphs in computers. The use of the adjacency matrices is, perhaps, one of the most common and convenient ways.

Exercise 1.2

(1) Let G be the multigraph representing the following diagram. Determine $V(G)$, $E(G)$, $v(G)$ and $e(G)$. Is G a simple graph?

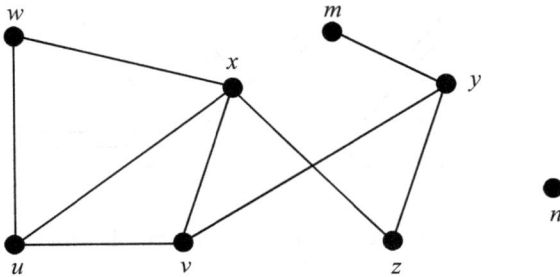

(2) Draw the graph G modeling the flight connectivity between twelve capital cities with the following vertex set $V(G)$ and edge set $E(G)$.

$V(G) = \{$Asuncion, Beijing, Canberra, Dili, Havana, Kuala Lumpur, London, Nairobi, Phnom Penh, Singapore, Wellington, Zagreb$\}$.

$E(G) = \{$Asuncion-London, Asuncion-Havana, Beijing-Canberra, Beijing-Kuala Lumpur, Beijing-London, Beijing-Singapore, Beijing-Phnom Penh, Dili-Kuala Lumpur, Dili-Singapore, Dili-Canberra, Havana-London, London-Wellington, Kuala Lumpur-London, Kuala Lumpur-Phnom Penh, Kuala Lumpur-Singapore, Kuala Lumpur-Wellington, London-Nairobi, Phnom Penh-Singapore, London-Singapore, London-Zagreb, Singapore-Wellington, Havana-Nairobi$\}$.

(Note that you may use A to represent Asuncion, B to represent Beijing, C to represent Canberra, etc.)

(3) Define a graph G such that $V(G) = \{2, 3, 4, 5, 11, 12, 13, 14\}$ and two vertices s and t are adjacent if and only if $gcd\{s, t\} = 1$. Draw a diagram of G and find its size $e(G)$.

(4) The diagram in page 12 is a map of the road system in a town. Draw a multigraph to model the road system, using a vertex to represent a junction and an edge to represent a road joining two junctions.

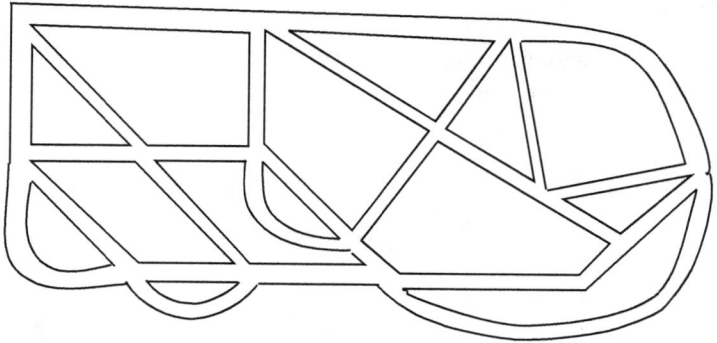

Diagram for Problem 4

(5) Let G be a graph with $V(G) = \{1, 2, \cdots, 10\}$, such that two numbers i and j in $V(G)$ are adjacent if and only if $|i - j| \leq 3$. Draw the graph G and determine $e(G)$.

(6) Let G be a graph with $V(G) = \{1, 2, \cdots, 10\}$, such that two numbers i and j in $V(G)$ are adjacent if and only if $i + j$ is a multiple of 4. Draw the graph G and determine $e(G)$.

(7) Let G be a graph with $V(G) = \{1, 2, \cdots, 10\}$, such that two numbers i and j in $V(G)$ are adjacent if and only if $i \times j$ is a multiple of 10. Draw the graph G and determine $e(G)$.

(8) Find the adjacency matrix of the following graph G.

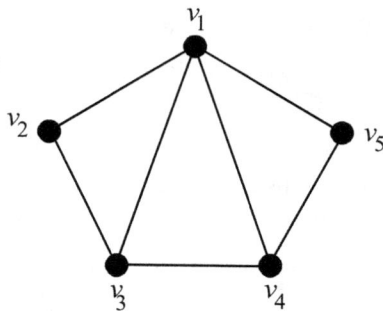

(9) The adjacency matrix of a multigraph G is shown below:

$$\begin{pmatrix} 0 & 1 & 0 & 2 & 3 \\ 1 & 0 & 1 & 2 & 2 \\ 0 & 1 & 0 & 1 & 1 \\ 2 & 2 & 1 & 0 & 1 \\ 3 & 2 & 1 & 1 & 0 \end{pmatrix}$$

Draw a diagram of G.

(10) Four teams of three specialist soldiers each (a scout, a signaler and a sniper) are to be sent into enemy territory. However, some of the soldiers cannot work well with some others. The following table shows the soldiers, their specializations and who they cannot work with.

Soldier	Specialization	Cannot cooperate with
1	Scout	$5, 7, 10$
2	Scout	—
3	Scout	$5, 6, 8, 9, 11$
4	Scout	$8, 12$
5	Signaler	$1, 3, 9$
6	Signaler	$3, 10, 11$
7	Signaler	$1, 9, 12$
8	Signaler	$3, 4, 9, 10$
9	Sniper	$3, 5, 7, 8$
10	Sniper	$1, 6, 8$
11	Sniper	$3, 6$
12	Sniper	$4, 7$

(i) Draw a multigraph to model the situation so that we may see how to form 3-man teams such that each specialization is represented and every member of the team can work with every other. State clearly what the vertices represent and under what condition(s) two vertices are joined by an edge.

(ii) Can you form four 3-man teams such that each specialization is represented and all members of the team can work with one another?

1.3   Vertex degrees

Let G be a multigraph.

Two vertices u and v in G are said to be **adjacent** if they are joined
by an edge, say, e in G. In the case when e is the only edge joining u
and v, we also write $e = uv$, and we say that
(1) u is a **neighbour** of v and vice versa,
(2) the edge e is **incident with** the vertex u (and v) and
(3) u and v are the two **ends** of e.

The set of all neighbours of v in G is denoted by $N(v)$; that is,

$$N(v) = \{x | x \text{ is a neighbour of } v\}.$$

Example 1.3.1.   *Let G be the multigraph shown in Figure 1.9.*

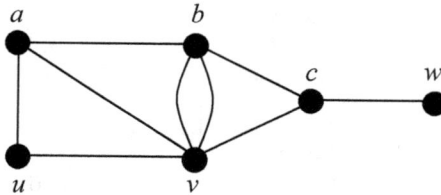

Figure 1.9

Then
(1) the vertices a and b are adjacent, so are b and v, but not a and c;
(2) the vertices a and u are the two ends of the edge au;
(3) the edge av is incident with the vertices a and v;
(4) the vertex a has three neighbours, namely, b, u and v; and
(5) $N(a) = \{b, u, v\}$, $N(b) = \{a, v, c\}$, $N(w) = \{c\}$, etc.

Let G be a multigraph. We now introduce a very useful and important
number associated with each vertex in G.

Given a vertex v in G, the **degree** of v in G, denoted by $d_G(v)$, is defined as the number of edges incident with v.

For simplicity, we shall replace $d_G(v)$ simply by $d(v)$ if there is no danger of confusion.

Question 1.3.1.

(i) Find the degree of each vertex in G of Figure 1.9.

(ii) Find $N(x)$ for each vertex x in G of Figure 1.9.

(iii) By definition, is it true that $d(v) = |N(v)|$?

Example 1.3.2. *Let G be the multigraph of Figure 1.10.*

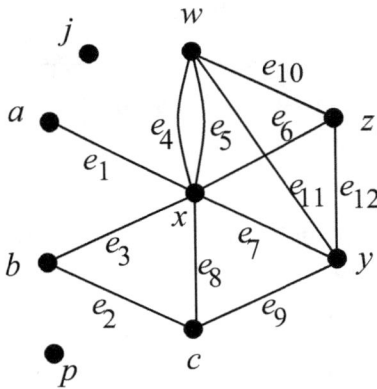

Figure 1.10

Observe that there are seven edges incident with the vertex x. Thus, $d(x) = 7$. There are no edges incident with the vertex j. Thus, $d(j) = 0$. The degrees of the vertices in G are shown in Table 1.11.

Vertex	a	b	c	j	p	w	x	y	z
Degree	1	2	3	0	0	4	7	4	3

Table 1.11

The degree of a vertex is also called the **valency** of the vertex as it is related to the valency of an atom in chemical compounds as shown in Figure 1.12.

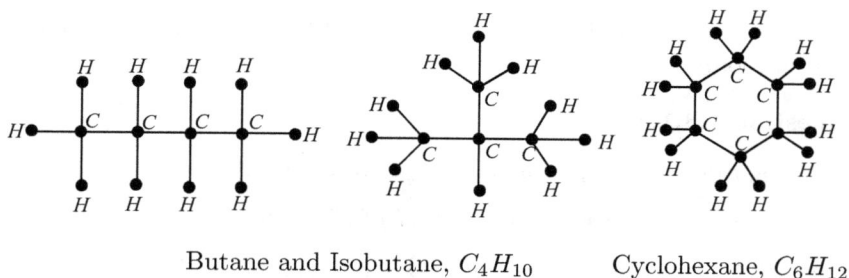

Butane and Isobutane, C_4H_{10} Cyclohexane, C_6H_{12}

Figure 1.12

Two types of vertices having smallest degrees have special names.

A vertex v is called an **isolated-vertex** if $d(v) = 0$; it is called an **end-vertex** if $d(v) = 1$.

Thus, in the multigraph of Figure 1.10, the vertices j and p are isolated-vertices and the vertex a is an end-vertex.

Remark. An *end-vertex* and an *end* of an edge (see page 14) are two different concepts. While an *end-vertex* is a vertex of degree one, an *end* of an edge has nothing to do with its degree.

1.3.1   *The Handshaking Lemma*

In the multigraph G of Figure 1.10, the total sum of the degrees of its vertices, as can be seen from Table 1.11, is 24. What is the size of G? The answer is: $e(G) = 12$. Observe that the total sum 24 is double the size 12 of G. Is this a coincidence?

Question 1.3.2. *Consider the multigraph G of Figure 1.9. Find e(G) and the sum of the degrees of the six vertices. Is the sum twice of e(G)?*

In general, is the sum of the degrees of the vertices in a multigraph always double its size?

There is a Chinese saying: *whenever you drink water, think of its source.* Where do the *degrees* of the vertices come from? The answer is: the *existence* of 'edges'. No edges implies no degrees. How many degrees can each edge contribute? The answer is '2' as an edge is incident with its two ends. *To compute the sum of degrees of vertices, we count each edge twice, once for each end of the edge.* Thus, we have the following result due to Euler [E]:

Theorem 1.1. *Let G be a multigraph with $V(G) = \{v_1, v_2, \cdots, v_n\}$. Then*

$$\sum_{i=1}^{n} d(v_i) = 2e(G).$$

Remarks. (1) If the vertices of G in Theorem 1.1 are not named as shown, the result can also be expressed as

$$\sum_{v \in V(G)} d(v) = 2e(G).$$

(2) A group of n (≥ 2) persons were together and various handshakes took place among them (two persons might even shake hands more than once). Each person recorded the number of handshakes he or she had shaken. The sum of these n numbers would be double the number of handshakes that took place in the gathering. Theorem 1.1 is, therefore, also known as the **Handshaking Lemma.**

An important way to classify the vertices of a multigraph G is by means of the **parity** (i.e., being even or odd) of their degrees.

A vertex w in G is said to be **even** if $d(w)$ is even; and said to be **odd** if $d(w)$ is odd.

Thus, in the multigraph G of Figure 1.10, there are five even vertices: b, j, p, w and y; and four odd vertices: a, c, x and z.

Question 1.3.3.   *(1) How many odd vertices are there in each of the multigraphs shown in the previous examples?*
(2) Can you construct a multigraph containing (i) exactly one odd vertex? (ii) exactly three odd vertices?

Instead of merely considering the multigraph of Figure 1.2, which represents the Königsberg bridge problem, Euler [E] studied a much more general problem: *Let G be a multigraph. Suppose that one starts with an arbitrary vertex in G, and finds that it is possible to have a walk which passes through each edge exactly once and then be able to end at the starting vertex. What can be said about such a multigraph?*

Evidently, in order to have such a walk one must be able to enter and exit a vertex; hence each vertex must be even. Is the converse true? We will discuss this in Chapter 6.

In order to study this problem and its related issues, Euler introduced the notion of 'odd vertices' and determined the parity of the number of 'odd vertices'. To get to this, Euler first established Theorem 1.1, and then deduced from it the following consequence:

Corollary 1.2. *The number of odd vertices in any multigraph is even.*

Proof.   Let G be a multigraph. Let A be the set of odd vertices in G, and B be the set of even vertices in G. Our aim is to show that $|A|$ is even. Indeed, as $V(G) = A \cup B$ and by Theorem 1.1, we have

$$\sum_{v \in A} d(v) + \sum_{v \in B} d(v) = \sum_{v \in V(G)} d(v) = 2e(G).$$

Since $\sum_{v \in B} d(v)$ and $2e(G)$ are even, $\sum_{v \in A} d(v)$ is also even. As $d(v)$ is odd for each v in A, it follows that $|A|$ must be even, as required. $\square$

Let us proceed to introduce two useful quantities pertaining to the degrees of vertices of a multigraph.

Let G be a multigraph. The **maximum degree** of G, denoted by $\Delta(G)$, is defined as the **maximum number** among all vertex degrees in G.

Likewise, the **minimum degree** of G, denoted by $\delta(G)$, is defined as the **minimum number** among all vertex degrees in G.

That is,

$$\Delta(G) = \max\{d(v)|v \in V(G)\} \text{ and}$$

$$\delta(G) = \min\{d(v)|v \in V(G)\}.$$

Thus, in the multigraph G of Figure 1.10, $\Delta(G) = 7$ and $\delta(G) = 0$.

Example 1.3.3. *Consider the graphs G and H shown in Figure 1.13 (a) and (b) respectively.*

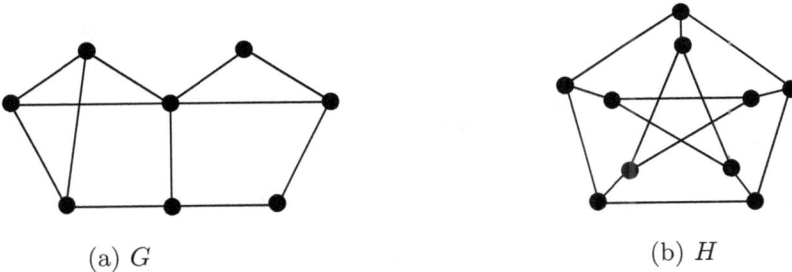

(a) G (b) H

Figure 1.13

It can be checked that $\Delta(G) = 5$ while $\delta(G) = 2$, and $\Delta(H) = \delta(H) = 3$.

Remark. The graph H shown in Figure 1.13(b) is a famous graph, known as the **Petersen graph**. It was named after Julius Petersen (1839-1910), a Danish mathematician, who discussed the graph in the paper [P].

1.3.2   *Regular graphs*

Notice that in the Petersen graph H, every vertex has the same degree, namely '3'. There are many graphs in which every vertex has the same degree. We now single out this special family of graphs by giving these graphs a name.

A graph G is said to be **regular** if every vertex in G has the same degree. More precisely, G is said to be k-**regular** if $d(v) = k$ for each vertex v in G, where $k \geq 0$.

Thus, a graph G is k-regular if and only if $\Delta(G) = \delta(G) = k$. Note that the Petersen graph is 3-regular. For $k = 0, 1, 2, 3$ and 4, a k-regular graph is shown in Figure 1.14.

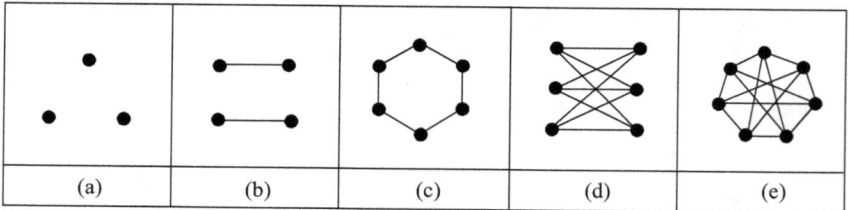

Figure 1.14

Remark. A 3-regular graph is also called a **cubic graph**.

Question 1.3.4.   *Construct a 5-regular graph of order* 10. *What is its size?*

To end this section, we introduce three important families of regular graphs: the '*null graphs*', '*complete graphs*' and '*cycles*'.

(1) The null graphs

By the definition of a graph G, the vertex set $V(G)$ is never empty, but its edge set $E(G)$ may be empty. The graph (a) in Figure 1.14 is an example.

A graph G is called a **null graph** (or **empty graph**) if $E(G)$ is empty. A null graph of order n is denoted by N_n.

Clearly, each N_n is a 0-regular graph and $e(N_n) = 0$. The five smallest null graphs are shown in Figure 1.15.

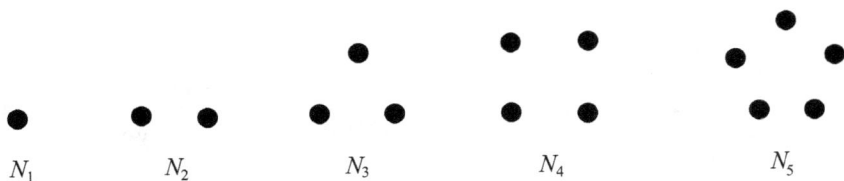

N_1 N_2 N_3 N_4 N_5

Figure 1.15

Among the (simple) graphs G of a fixed order n, at one extreme, the null graph N_n contains no edges (the least possible size). At the other extreme, we may ask:

Question 1.3.5. *What is the largest possible size that G can have? Which graph has its size attaining this largest possible number?*

(2) **The complete graphs**

While a null graph is one in which no two vertices are adjacent, a graph is called a **complete graph** if every two of its vertices are adjacent. A complete graph of order n is denoted by K_n.

Clearly, each K_n is $(n-1)$-regular and $e(K_n) = \binom{n}{2}$. The first five smallest complete graphs are shown in Figure 1.16.

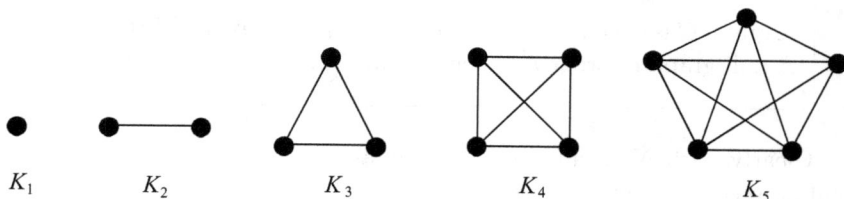

Figure 1.16

(3) The cycles

The 2-regular graph shown in Figure 1.14(c) is called a **cycle**.

> A graph G of order $n \geq 3$ is called a **cycle** if its n vertices can be named as $v_1, v_2, \cdots, v_n$ such that v_1 is adjacent to v_2, v_2 is adjacent to v_3, $\cdots$, v_{n-1} is adjacent to v_n, v_n is adjacent to v_1, and no other adjacency exists; that is,
>
> $$V(G) = \{v_1, v_2, \cdots, v_n\} \text{ and}$$
> $$E(G) = \{v_1v_2, v_2v_3, \cdots, v_{n-1}v_n, v_nv_1\}.$$
>
> A cycle of order n is denoted by C_n. We call C_n an n**-cycle**, and C_3 a **triangle**.

Clearly, every cycle is 2-regular and $e(C_n) = v(C_n) = n$. Three more examples of cycles are shown in Figure 1.17.

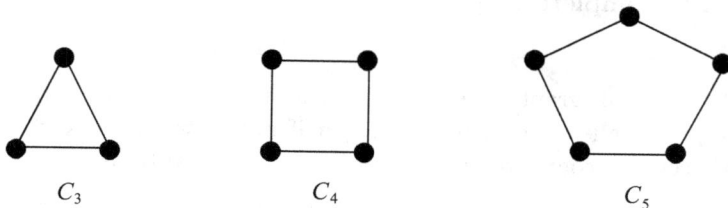

C_3 C_4 C_5

Figure 1.17

Remark. The graph C_n is defined for $n \geq 3$. For $n = 2$, as shown in Figure 1.18, C_2 is also called a cycle, but it is not a graph (it is a multigraph).

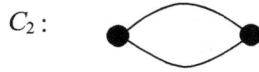

$C_2:$

Figure 1.18

Exercise 1.3

(1) In the following multigraph G, find

    (i) the size of G,
    (ii) the degree of each vertex,
    (iii) the sum $\sum\{d(v)|v \in V(G)\}$,
    (iv) the number of odd vertices,
    (v) $\Delta(G)$, and
    (vi) $\delta(G)$.

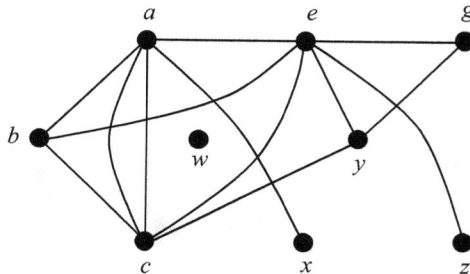

Is your answer for (iii) double your answer for (i)? Is your answer for (iv) an even number?

(2) Construct a multigraph of order 6 and size 7 in which every vertex is odd.

(3) Let G be a multigraph with $V(G) = \{v_1, v_2, \cdots, v_n\}$. Prove that the sum of all the entries in the ith row of the adjacency matrix $A(G)$ is the degree of the vertex v_i for each $i = 1, 2, \cdots, n$.

(4) Let G be a graph of order 8 and size 15 in which each vertex is of degree 3 or 5. How many vertices of degree 5 does G have? Construct one such graph G.

(5) Let H be a graph of order 10 such that $3 \le d(v) \le 5$ for each vertex v in H. Not every vertex is even. No two odd vertices are of the same degree. What is the size of H?

(6) Let G be a graph of order 14 and size 30 in which every vertex is of degree 4 or 5. How many vertices of degree 5 does G have? Construct one such graph G.

(7) Does there exist a multigraph G of order 8 such that $\delta(G) = 0$ while $\Delta(G) = 7$? What if 'multigraph G' is replaced by 'graph G'?

(8) Characterize the 1-regular graphs.

(9) Draw all regular graphs of order n, where $2 \le n \le 6$.

(10) (i) Does there exist a graph G of order 5 such that $\delta(G) = 1$ and $\Delta(G) = 4$?
   (ii) Does there exist a graph G of order 5 which has two vertices of degree 4 and $\delta(G) = 1$?

(11) Let H be a graph of order 8 and size 13 with $\delta(H) = 2$ and $\Delta(H) = 4$. Denote by n_i the number of vertices in H of degree i, where $i = 2, 3, 4$. Assume that $n_3 \ge 1$. Find all possible answers for (n_2, n_3, n_4). For each of your answers, construct a corresponding graph.

(12) Suppose G is a k-regular graph of order n and size m, where $k \ge 0$, $m \ge 0$ and $n \ge 1$. Find a relation linking k, n and m. Justify your answer.

(13) Does there exist a 3-regular graph with eight vertices? Does there exist a 3-regular graph with nine vertices?

(14) Construct a cubic graph of order 12. What is its size? Does there exist a cubic graph of order 11? Why?

(15) Let H be a k-regular graph of order n. If $e(H) = 10$, find all possible values for k and n; and for each case, construct one such graph H.

(16) (+) Let G be a 3-regular graph with $e(G) = 2v(G) - 3$. Determine the values of $v(G)$ and $e(G)$. Construct all such graphs G.

(17) Find all integers n such that $100 \le e(K_n) \le 200$.

(18) (+) Let G be a multigraph of order 13 in which each vertex is of degree 7 or 8. Show that G contains **at least eight** vertices of degree 7 or **at least seven vertices** of degree 8.

(19) (+) Let G be a graph of order n in which there exist **no** three vertices u, v and w such that uv, vw and wu are all edges in G. Show that $n \ge \delta(G) + \Delta(G)$.

(20) (+) There were n (≥ 2) persons at a party and, as usually happens, some shake hands with others. No one shook hands with the same person more than once. Show that there are at least two persons in the party who had the same number of handshakes.

(21) The preceding problem says that in any graph of order $n \geq 2$, there exist two vertices having the same degree. Is the result still valid for multigraphs?

(22) (+) Mr. and Mrs. Samy attended an exclusive party where in addition to themselves, there were only another 3 couples. As usually happens, some shake hands with others. No one shook hands with the same person more than once and no one shook hands with his/her spouse. After all the handshakes had been done, Mr. Samy asked each person, including his wife, how many hands he/she had shaken. To everyone's amusement, each one gave a different answer. How many hands did Mrs. Samy shake?

(23) (+) In the preceding problem, there were four couples altogether in a party. Solve the general problem where 'four couples' is replaced by 'n (≥ 2) couples'.

(24) (+) There are $n \geq 2$ distinct points in the plane such that the distance between any 2 points is at least one. Prove that there are at most $3n$ pairs of these points at distance exactly one.

1.4   Paths, cycles and connectedness

Figure 1.19(a) shows a section of the street system of a town. It can be modeled as a graph as shown in Figure 1.19(b), where a vertex represents a junction and two vertices are joined by an edge if and only if the corresponding junctions are linked by a street. For certain purposes, we may have to traverse the street system by passing through some junctions and streets. In order to show more precisely and succinctly the way we traverse, in this section, we shall introduce some basic terms in general multigraphs which serve the purpose.

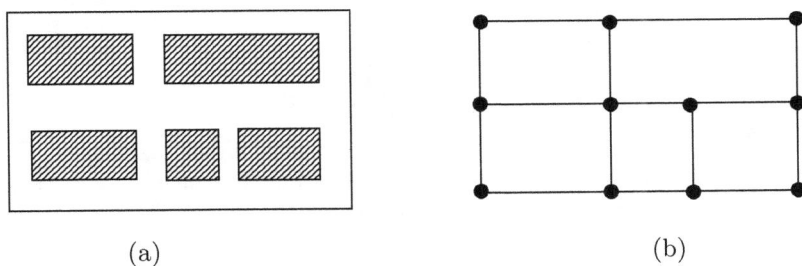

(a)   (b)

Figure 1.19

1.4.1   *Walks, trails, paths and cycles*

Consider the multigraph H of Figure 1.20. If we start at vertex a, then we can reach vertex x via the edge f_1, and from x to y via the edge f_7. We can further proceed to reach z via f_{12}. This process can be conveniently expressed by the following alternating sequence of vertices and edges:

$$a \ f_1 \ x \ f_7 \ y \ f_{12} \ z.$$

Such a sequence is called a **walk** or, more precisely, an $a - z$ walk as a and z are, respectively, the **initial** and **terminal** vertices of the walk.

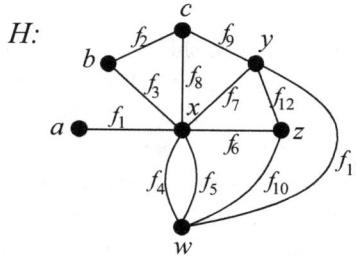

Figure 1.20

A **walk** in a multigraph G is an alternating sequence of vertices and edges beginning and ending at vertices:

$(\#)$ $\qquad\qquad\qquad v_0 e_0 v_1 e_1 v_2 \cdots v_{k-1} e_{k-1} v_k,$

where $k \geq 1$ and e_i is incident with v_i and v_{i+1}, for each $i = 0, 1, \cdots, k-1$. The walk $(\#)$ is also called a $v_0 - v_k$ **walk** with its **initial vertex** v_0 and **terminal vertex** v_k. The **length** of the walk $(\#)$ is defined as 'k', which is the number of occurrences of edges in the sequence.

Remark. The vertices v_i's or edges e_i's in $(\#)$ need not be distinct.

Question 1.4.1. *Is the sequence: $c f_9 y f_{11} w f_1 b$ a walk in H of Figure 1.20? Why?*

Example 1.4.1. *Some walks in H of Figure 1.20 and their respective lengths are shown in Table 1.21.*

	sequence	walk	length
(1)	$bf_3xf_4wf_5xf_4wf_{10}zf_{10}w$	$b-w$	6
(2)	$bf_3xf_4wf_5xf_7y$	$b-y$	4
(3)	$bf_3xf_4wf_{11}yf_9c$	$b-c$	4
(4)	$bf_2cf_8xf_4wf_5xf_3b$	$b-b$	5
(5)	$bf_2cf_9yf_{12}zf_6xf_3b$	$b-b$	5

Table 1.21

The definition of a *walk* is quite general. In certain circumstances, we will need some special types of walks. A highway inspector may not want to inspect a road twice. And a traveller may not want to visit a city more than once.

A walk is called a **trail** if no *edge* in it is traversed more than once. A walk is called a **path** if no *vertex* in it is visited more than once.

Question 1.4.2. *Is every trail a path? Is every path a trail?*

In Example 1.4.1,
walk (1) is neither a trail nor a path (why?);
walk (2) is a trail but not a path (why?);
walk (3) is both a trail and a path (why?);
walks (4) and (5) are both trails but not paths (why?).

Remark. It is now clear that if some edge is traversed more than once, then at least one of its two ends is visited more than once. Thus, **every path must be a trail**. Its converse is, however, not true.

The walk (1) in Example 1.4.1 is a $b-w$ walk, but not a $b-w$ path as the vertices x and w are visited more than once. However, it can be cut short, say to bf_3xf_4w, to become a $b-w$ path. Likewise, the $b-y$ walk is not a $b-y$ path, but it can be cut short to bf_3xf_7y to become a $b-y$ path.

Question 1.4.3. *Is it true that every $u-v$ walk always contains a $u-v$ path?*

A $u - v$ walk is said to be **closed** if $u = v$, that is, its initial and terminal vertices are the same; and **open** otherwise.

Thus, in Example 1.4.1, the walks (1), (2) and (3) are open while the walks (4) and (5) are closed.

A closed walk of length at least two in which no edge is repeated is called a **circuit**.

Thus, in Example 1.4.1, the closed walks (4) and (5) are circuits. Note that vertices are allowed to be repeated in a circuit.

Question 1.4.4.   *In the multigraph H of Figure 1.20, find a circuit of length 2 and a circuit of length 8.*

A circuit is called a **cycle** if no vertex is repeated (except the initial and terminal vertices).

Thus, in Example 1.4.1, while the circuit (4) is not a cycle (why?), the circuit (5) is a cycle.

Question 1.4.5.   *For each $k = 3, 4, 5, 6$, find a cycle of length k which passes through the vertex z in the multigraph of Figure 1.20.*

Question 1.4.6.   *The circuit (4) in Example 1.4.1 is not a cycle, but it can be cut short to $bf_2cf_8xf_3b$ to become a cycle. Is it true that every circuit always contains a cycle?*

Remark. At the end of Section 1.3, a cycle is introduced as a graph. In the above discussion, however, a cycle is regarded as a special closed walk in a multigraph. What a cycle means should be clear from the context when it is used.

Remark. We express a walk in a multigraph as an alternating sequence of vertices and edges. It is necessary to name the edges since two vertices may be joined by more than one edge and we want to know which edge is traversed to visit these two vertices. However, when we confine ourselves to graphs, as two adjacent vertices are joined by a unique edge, such an expression can be simplified by dropping the names of edges. Thus, in the graph G of Figure 1.22, the $u - w$ walk $ue_1ve_7ye_4ze_3w$ can simply be denoted by $uvyzw$ without any ambiguity.

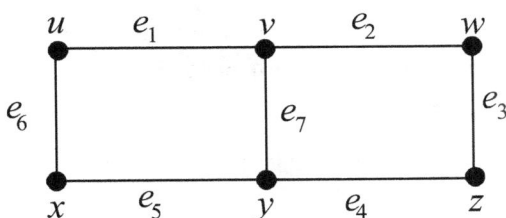

Figure 1.22

1.4.2 *Connected multigraphs*

The notion of walks or paths enables us to introduce a very important class of multigraphs, called **connected** multigraphs.

> A multigraph G is said to be **connected** if every two vertices in G are joined by a path.

Example 1.4.2. *There are two graphs G and H in Figure 1.23. It can easily be checked that every two vertices in G are joined by a path. Thus G is a connected graph. However, the graph H is not connected since, for instance, the vertices r and u in H are not joined by any path.*

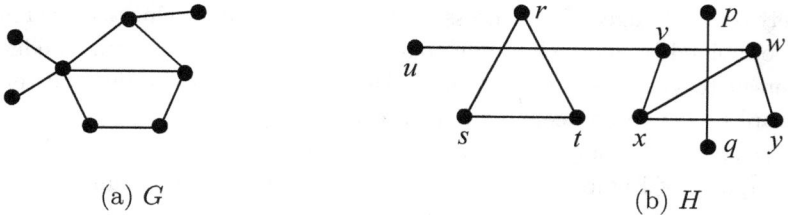

(a) *G*   (b) *H*

Figure 1.23

A multigraph is said to be **disconnected** if it is not connected.

Thus the graph *H* in Figure 1.23(b) is disconnected.

Question 1.4.7.   *Consider the disconnected graph H in Figure 1.23(b).*
(1) Which vertices are reachable from the vertex r via a path?
(2) Which vertices are reachable from the vertex u via a path?
(3) Which vertices are reachable from the vertex p via a path?

The answers to (1), (2) and (3) of the above question are, respectively, shown in Figure 1.24. Note that each of them is a connected 'piece', and is called a **connected component** of *H*. From now on, we simply call a connected component a **component**. Thus the disconnected graph *H* has three components.

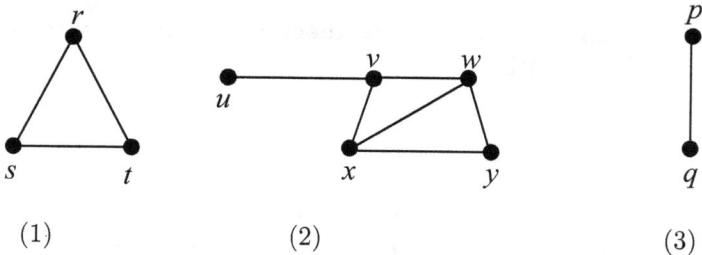

(1)   (2)   (3)

Figure 1.24

Question 1.4.8. *Consider the graph of order 12 and size 9 in Figure 1.25. Is the graph connected? If not, how many components does it have?*

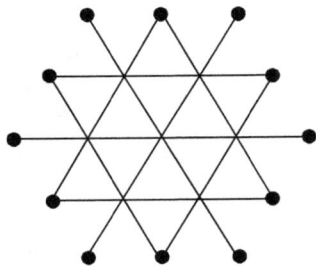

Figure 1.25

1.4.3 Distance

To end this chapter, we introduce in what follows an important quantity associated with a pair of vertices in a connected multigraph.

The graph H shown in Figure 1.26(a) is connected, and any two vertices in H are joined by at least one path. For instance, some paths joining the vertices x and w and their respective lengths are shown in Figure 1.26(b).

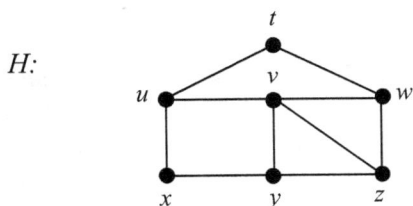

$x - w$ path	length
$xyzvutw$	6
$xuvyzw$	5
$xyzvw$	4
$xutw$	3
$xyzw$	3

(a)   (b)

Figure 1.26

We notice that '3' is the smallest length in the list in Figure 1.26(b). We ask: is there any $x - w$ path of length less than '3' in H? The answer is 'no'. That is, among all paths joining x and w in H, the smallest length is '3'. In this case, we say that the **distance from** x **to** w **is** '3', and we write $d(x, w) = 3$.

Let G be a connected multigraph, and u, v be any two vertices in G. The **distance from** u **to** v, denoted by $d(u, v)$, is defined as the *smallest length* of all $u - v$ paths in G.

Question 1.4.9.   *In the graph H of Figure 1.26(a), find $d(u, u), d(x, y)$, $d(y, x)$, $d(x, z), d(z, x), d(u, z)$, and $d(t, y)$.*

Let G be a connected multigraph, and x, y, z be any vertices in G. Some facts on distances are stated below.

(1) $d(x, x) = 0$,
(2) $d(x, y) > 0$ if $x \neq y$,
(3) $d(x, y) = d(y, x)$,
(4) $d(x, y) + d(y, z) \geq d(x, z)$.

Question 1.4.10.   *(1) If $d(x, y) = 1$, what is the relation between x and y?*

(2) If $d(x, y) > 1$, what is the relation between x and y?
(3) Is it possible that $d(x, y) + d(y, z) = d(x, z)$?
(4) Is it possible that $d(x, y) + d(y, z) > d(x, z)$?

The greatest distance between any two vertices in a graph G (i.e. $\max\{d(u, v)|u, v \in V(G)\}$) is called the **diameter** of G. For instance, the diameter of the graph of Figure 1.26 (a) is '3'. Just as the diameter of a circle is the greatest distance between any two points on the circle, the diameter of a graph is an indication of how 'far apart' vertices in a graph are.

Suppose we represent each person on earth by a vertex and put an edge between two vertices if the two persons are acquainted. What will be the diameter of this 'acquaintance' graph? In 1990, John Guare wrote a play called 'Six Degrees of Separation' in which it is claimed that any two persons can be connected by a chain of at most 6 intermediaries, i.e. the diameter of the acquaintance graph is 6. This claim was a result of a social experiment by Stanley Milgram. He gave some volunteers in Nebraska and

Kansas (both Midwestern states of the USA) packages with the name of an individual in Massachusetts on the east coast of the USA. Each volunteer was asked to pass the package on to an acquaintance who he thought could get the package 'nearer' to the intended recipient, with the instruction to pass it on to another acquaintance, and so on, till it finally reaches the intended recipient. The average number of intermediaries for the packages that finally made it to the recipient was 5.5!

Does this prove Guare's claim? Certainly not, since the packages that did not reach their destination could not be counted, and also those who did may not have taken the shortest 'path'. In addition, the experiment was only within the USA and there are many parts of the world that are clearly more isolated than Nebraska or Kansas. Still, the idea of a 'small world' seems plausible and the diameter of the acquaintance graph may indeed be a small number.

Exercise 1.4

(1) Consider the following graph H.

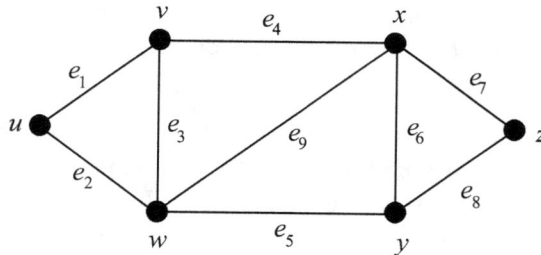

    (a) Which of the following sequences represents a $u - z$ walk in H?
       (i) $ue_2we_5xe_7z$
       (ii) $ue_1ve_5ye_8z$
       (iii) $ue_1ve_3we_3ve_4xe_7z$
    (b) Find a $u - z$ trail in H that is not a path.
    (c) Find all $u - z$ paths in H which pass through e_9.

(2) Consider the following multigraph G:

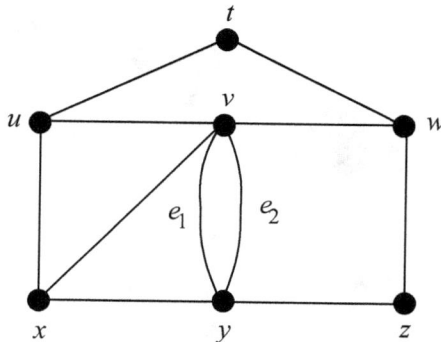

    (a) Find $d(t,v)$, $d(t,y)$, $d(x,w)$ and $d(u,z)$.
    (b) For $k = 2, 3, 4, 5, 6, 7$, find a cycle of length k in G.
    (c) Find a circuit of length 6 in G that is not a cycle.
    (d) Find a circuit of length 8 in G that does not contain t.
    (e) Find a circuit of length 9 in G that contains t and v.

(3) Is the following graph H disconnected? If it is so, find its number of components.

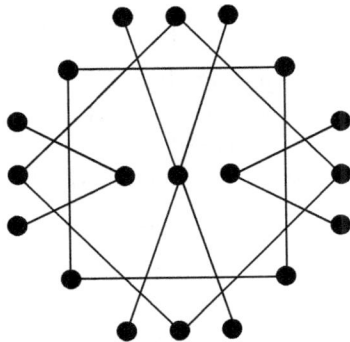

(4) Let G be a graph with $V(G) = \{1, 2, \cdots, n\}$, where $n \geq 5$, such that two numbers i and j in $V(G)$ are adjacent if and only if $|i - j| = 5$. How many components does G have?

(5) (+) Show that any $u - v$ walk in a graph contains a $u - v$ path.

(6) (+) Show that any circuit in a graph contains a cycle.

(7) (+) Show that any graph G with $\delta(G) \geq k$ has a path of length k.

(8) (+) Let G be a graph of order $n \geq 2$ such that $\delta(G) \geq \frac{1}{2}(n-1)$. Show that $d(u, v) \leq 2$ for any two vertices u, v in G.

(9) (+) Let G be a graph of order n and size m such that $m > \binom{n-1}{2}$. Show that G is connected.

(10) For $n \geq 2$, construct a disconnected graph of order n and size $\binom{n-1}{2}$.

(11) Let G be a disconnected graph of order 5. What is the largest possible value for $e(G)$? If G is a disconnected graph of order $n \geq 2$, what is the largest possible value for $e(G)$? Construct one such extremal graph of order n.

(12) (+) Let G be a graph of order $n \geq 2$ and u, v be two non-adjacent vertices in G such that $d(u) + d(v) \geq n + r - 2$. Show that u and v have at least r common neighbours.

(13) (+) Let G be a connected graph that is not complete. Show that there exist three vertices x, y, z in G such that x and y, y and z are adjacent, but x and z are not adjacent in G.

(14) (+) Let G be a graph of order n and size m such that $\Delta(G) = n - 2$ and $d(u, v) \leq 2$ for any two vertices u, v in G. Show that $m \geq 2n - 4$.

(15) Let G be a graph such that $N(x) \cup N(y) = V(G)$ for every pair of vertices x, y in G. What can be said of G?

(16) (+) Let H be a graph of order $n \geq 2$. Suppose that H contains two distinct vertices u, v such that (i) $N(u) \cup N(v) = V(H)$ and (ii) $N(u) \cap N(v)$ is non-empty.
What is the least possible value of $e(H)$?

(17) Suppose G is a disconnected graph which contains exactly two odd vertices u and v. Must u and v be in the same component of G? Why?

(18) (+) Show that any two longest paths in a connected graph have a vertex in common.

(19) (+) Show that a graph G is connected if and only if for any partition of $V(G)$ into two non-empty sets A and B, there is an edge in G joining a vertex in A and a vertex in B.

(20) (+) Suppose G is a connected graph with k edges. Prove that it is possible to label the edges $1, 2, \cdots, k$ in such a way that at each vertex which belongs to two or more edges (i.e. which is of degree at least two), the greatest common divisor of the integers labeling those edges is 1 (32nd IMO, 1991/4).

Chapter 2

Graph Isomorphisms, Subgraphs, the Complement of a Graph

2.1 Isomorphic graphs and isomorphisms

Consider the following three quadrilaterals:

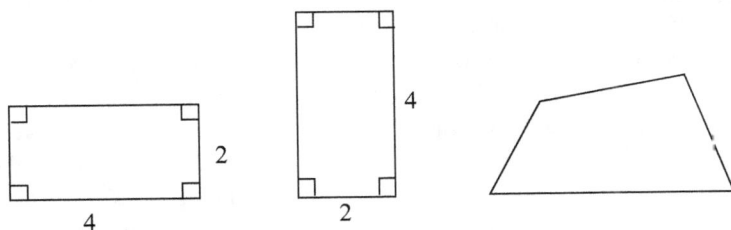

In **plane geometry**, we would say that the first two are the 'same' (i.e., **congruent**), and that both of them are 'different' from the third one.

Suppose now we are asked to draw a graph G which is defined as follows: its vertex set $V(G) = \{w, x, y, z\}$ and edge set $E(G) = \{wx, xy, yz, zw\}$. Some of us may place the four vertices as shown in Figure 2.1(a), others may place them as shown in Figure 2.1(b), (c), etc.

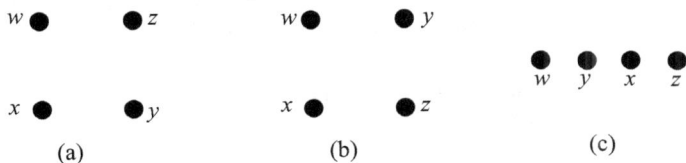

Figure 2.1

39

By joining some four pairs of vertices with the four edges as given in $E(G)$, we would have their corresponding diagrams as shown in Figure 2.2.

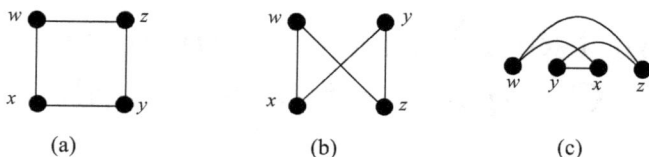

(a)   (b)   (c)

Figure 2.2

Apparently, these three diagrams look very different 'geometrically'. However, in the context of 'graphs', they are absolutely the 'same'.

In the study of an area of mathematics (such as plane geometry and graph theory), the first thing to do before proceeding any further is to know whether two objects under consideration (such as quadrilaterals in plane geometry and graphs in graph theory) are the same or are different.

Intuitively, two graphs G and H are considered the 'same' if it is possible to relocate the vertices of one of the graphs, say G, so that these vertices have the same positions as the vertices in H, the result of which is that the two graphs look identical (imagine that the edges are rubber bands; see Figure 2.3). Mathematically, we use a more fancy term, **isomorphic graphs**, to replace 'same graphs' and define it as follows:

Two graphs G and H are said to be **isomorphic** if there exists a one-one and onto mapping $f : V(G) \longrightarrow V(H)$ such that two vertices u, v are adjacent in G when and only when their images $f(u)$ and $f(v)$ under f are adjacent in H (i.e., **the adjacency is preserved under f**).

In this case, we shall write $G \cong H$ and call the mapping f an **isomorphism** from G to H.

Remark. The word **isomorphism** is derived from the Greek words *isos* (meaning 'equal') and *morphe* (meaning 'form').

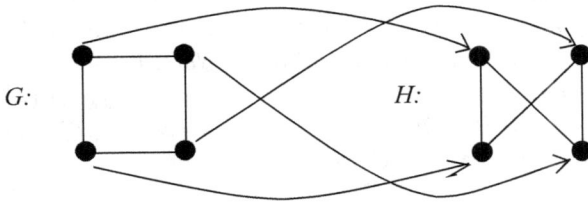

Figure 2.3

Example 2.1.1. *Consider the graphs G and H as shown in Figure 2.4. We claim that $G \cong H$. Indeed, if we define a mapping $f : V(G) \longrightarrow V(H)$ by $f(v_i) = u_i$ for each $i = 1, 2, \cdots, 6$, then it can be checked that f is both one-one and onto, and that the adjacency is preserved under f. Thus G and H are isomorphic under the isomorphism f.*

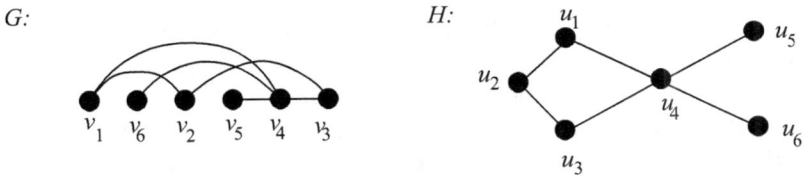

Figure 2.4

Question 2.1.1. *Let G and H be the graphs of Figure 2.4.*

(1) *Define $g : V(G) \to V(H)$ by $g(v_2) = u_3$, $g(v_3) = u_2$, $g(v_i) = u_i$ for $i = 1, 4, 5, 6$. Does g preserve the adjacency? Is g an isomorphism?*

(2) *Define $h : V(G) \to V(H)$ by $h(v_1) = u_3$, $h(v_3) = u_1$, $h(v_5) = u_6$, $h(v_6) = u_5$, $h(v_i) = u_i$ for $i = 2, 4$. Does h preserve the adjacency? Is h an isomorphism?*

Remark. Given that two graphs are isomorphic, there often exist more than one isomorphism from one of the graphs to the other. We have shown two isomorphisms from G to H (namely, f and h) of Figure 2.4. The reader is encouraged to find another isomorphism from G to H.

Example 2.1.2.   *Consider the graphs G and H as shown in Figure 2.5. Define a mapping $f : V(G) \longrightarrow V(H)$ by $f(x_i) = y_i$ for each $i = 1, 2, 3, 4$. It is clear that f is both one-one and onto. Note, however, that x_2 and x_4 are adjacent in G but their images $f(x_2) (= y_2)$ and $f(x_4) (= y_4)$ under f are not adjacent in H. Thus f does not preserve the adjacency, and so f is* **not** *an isomorphism from G to H.*

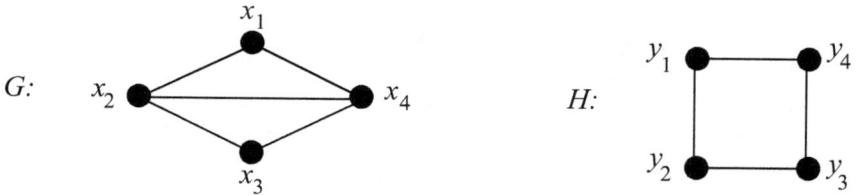

Figure 2.5

Example 2.1.3.   *Consider the graphs G and H as shown in Figure 2.6. Define a mapping $f : V(G) \longrightarrow V(H)$ by $f(a_i) = b_i$ for each $i = 1, 2, 3, 4$. It is obvious that f is both one-one and onto. Observe, however, that a_2 and a_4 are not adjacent in G but their images $f(a_2) (= b_2)$ and $f(a_4) (= b_4)$ are adjacent in H. Thus f does not preserve the adjacency, and so f is* **not** *an isomorphism from G to H.*

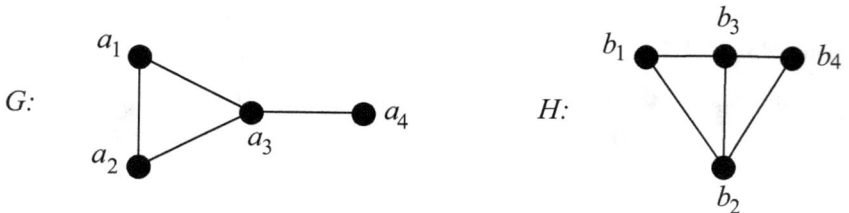

Figure 2.6

Example 2.1.4.   *Consider the graphs G and H as shown in Figure 2.7 and define a mapping $f : V(G) \longrightarrow V(H)$ by $f(w_i) = z_i$ for each $i =$*

$1, 2, \cdots, 5$. *Though f is both one-one and onto, it is clear that f does not preserve the adjacency, and so f is not an isomorphism from G to H. However, it does not mean that G is not isomorphic to H. Indeed, $G \cong H$ and the mapping $g : V(G) \longrightarrow V(H)$, defined by $g(w_1) = z_4$, $g(w_2) = z_2$, $g(w_3) = z_5$, $g(w_4) = z_3$ and $g(w_5) = z_1$, is an isomorphism from G to H.*

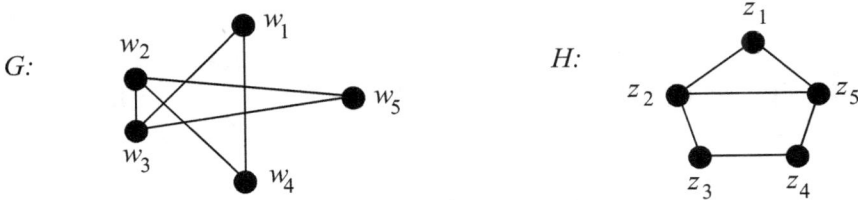

Figure 2.7

Question 2.1.2. *Suppose f is an isomorphism from a graph G to a graph H. Let u, v, w be three vertices in G. Assume that u, v, w form a K_3 in G (i.e., any two of them are adjacent in G). Do $f(u), f(v)$ and $f(w)$ also form a K_3 in H?*

Question 2.1.3. *Let F, G and H be any graphs. Is it true that:*
  (i) $G \cong G$?
  (ii) if $G \cong H$, then $H \cong G$?
  (iii) if $F \cong G$ and $G \cong H$, then $F \cong H$?

Question 2.1.4. *We have defined the concept of an isomorphism between two graphs. Could it be generalized to multigraphs? Are the multigraphs of Figures 1.2 and 1.3 the 'same'?*

2.2   Testing isomorphic graphs

To show that two given graphs are isomorphic, all we need is to find an isomorphism between them as done in Examples 2.1.1 and 2.1.4. How about claiming that two given graphs are not isomorphic? Could we simply say that 'it is so because *there is no isomorphism between them*'? This argument is certainly not convincing in general unless we do list all the one-

one and onto mappings between the two vertex sets (which are, however, too many if the orders of the graphs considered are large), and verify that none of them preserves the adjacency.

Recall that two graphs are isomorphic if we can find a one-one and onto mapping between their vertex sets which preserves the adjacency. It thus follows readily that if $G \cong H$, then G and H must have the same order and same size. That is:

Result (1)   If $G \cong H$, then $v(G) = v(H)$ and $e(G) = e(H)$.

Write $G \ncong H$ if the graphs G and H are not isomorphic. Then, equivalently, Result (1) says that if $v(G) \neq v(H)$ **or** $e(G) \neq e(H)$, then $G \ncong H$. As an application of this observation, we see readily that the graphs G and H in Example 2.1.2 (resp., Example 2.1.3) are not isomorphic.

Question 2.2.1.   *Are K_3 and C_3 isomorphic? Are K_4 and C_4 isomorphic?*

Question 2.2.2.   *Among the four graphs given in Figure 2.8, are there any two which are isomorphic?*

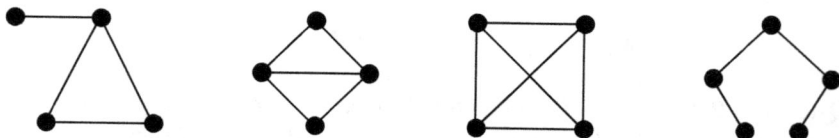

Figure 2.8

Recall that the degree $d(v)$ of a vertex v in a graph G is the number of edges incident with it. Assume that $V(G) = \{u_1, u_2, \cdots, u_n\}$. Call the sequence $(d(u_1), d(u_2), \cdots, d(u_n))$ the **degree sequence** of G. We may rename the vertices in G so that $d(u_1) \geq d(u_2) \geq \cdots \geq d(u_n)$. For instance, in the graph G of Figure 2.9, the five vertices are named as $u_1, \cdots, u_5$ so that the degree sequence of G is given by $(3, 2, 2, 2, 1)$, which is in non-increasing order.

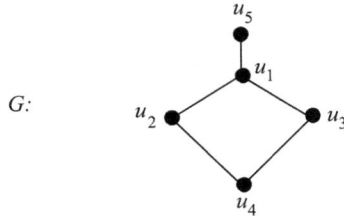

G:

Figure 2.9

Suppose that two graphs G and H are isomorphic under an isomorphism f. As f preserves the adjacency, it follows that, for each vertex v in G, $d(v) = d(f(v))$ (see Problem 7 of Exercise 2.2). Thus, we have:

Result (2). If $G \cong H$, then G and H have the same degree sequence, in non-increasing order.

Hence, equivalently, if G and H have different degree sequences in non-increasing order, then $G \not\cong H$. As an application of this fact, let us consider the following:

Example 2.2.1. *Determine whether the graphs of Figure 2.10 are isomorphic:*

G: H:

Figure 2.10

The degree sequence of G is $(3, 2, 2, 2, 1)$ while that of H is $(2, 2, 2, 2, 2)$, which are different. Thus, $G \not\cong H$.

Remark. Actually, in Example 2.2.1, we do not need to use that 'big' notion of the degree sequence to conclude that $G \not\cong H$. We could arrive at the same by simply pointing out a simple observation that G has an end-vertex (a vertex of degree 1) while H does not have.

Question 2.2.3. *The graphs G and H in Example 2.2.1 are of the same order and same size, yet they are not isomorphic. This shows that the converse of Result (1) is false. Does the converse of Result (2) hold? Could you find two graphs that are not isomorphic but have the same degree sequence, in non-increasing order?*

Given two arbitrary graphs G and H of the same order and same size, is there an 'efficient' procedure which enables us to determine whether $G \cong H$? This problem, known as the **Isomorphism Testing Problem**, is a very difficult problem, and until now, only little progress has been made. There are many practical applications which desire a fast procedure to test graph isomorphism. For example, organic chemists who routinely deal with graphs which represent molecular links would like some system to quickly give each graph a unique name. Thus, many research papers have been published which discuss how to build fast and practical isomorphism testers.

For a good survey paper on the Graph Isomorphism Problem, the reader may refer to the paper by Fortin [F] or the book by Kobler et al. [KST].

Adjacency Matrices and Isomorphism

Recall from Chapter 1 that a graph can be represented by its adjacency matrix. We note first that the adjacency matrix of a graph depends on the ordering of the vertices of the graph. Consider the following graph G:

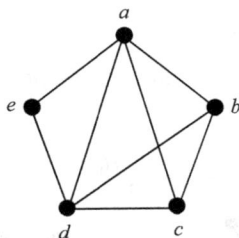

If we take the vertices in the ordering a, b, c, d, e, then the adjacency matrix of G is:

$$
\begin{array}{c}
\quad\; a\; b\; c\; d\; e \\
\begin{array}{c} a \\ b \\ c \\ d \\ e \end{array}
\left(\begin{array}{ccccc}
0 & 1 & 1 & 1 & 1 \\
1 & 0 & 1 & 1 & 0 \\
1 & 1 & 0 & 1 & 0 \\
1 & 1 & 1 & 0 & 1 \\
1 & 0 & 0 & 1 & 0
\end{array}\right)
\end{array}
$$

However, if we take the vertices in the ordering a, d, b, c, e, then the adjacency matrix is:

$$
\begin{array}{c}
\quad\; a\; d\; b\; c\; e \\
\begin{array}{c} a \\ d \\ b \\ c \\ e \end{array}
\left(\begin{array}{ccccc}
0 & 1 & 1 & 1 & 1 \\
1 & 0 & 1 & 1 & 1 \\
1 & 1 & 0 & 1 & 0 \\
1 & 1 & 1 & 0 & 0 \\
1 & 1 & 0 & 0 & 0
\end{array}\right)
\end{array}
$$

Hence, a graph of order n can have up to $n!$ adjacency matrices. On the other hand, if two graphs G and H possess identical adjacency matrices, we can conclude that G and H are isomorphic. This is true because we can immediately find an isomorphism between $V(G)$ and $V(H)$ by mapping the vertices according to their order in each adjacency matrix. (The reader may like to try to produce a more formal proof in Problem 12 of Exercise 2.2.)

Because it is difficult to find an isomorphism between two (isomorphic) graphs directly from their drawings, it may be easier to show that the graphs are isomorphic from the respective adjacency matrices instead. The initial adjacency matrices may not be the same but subsequent 'intelligent' reordering of the vertices of one adjacency matrix may result in the adjacency matrix of the other graph. In fact, a computer cannot find an isomorphism based on the drawings of two graphs and need inputs in the nature of adjacency matrices.

Example 2.2.2.   *Check if the following graphs G and H are isomorphic.*

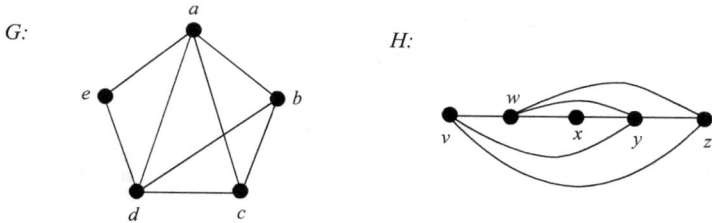

The adjacency matrices for G and H in the natural lexicographic ordering of their vertices are as follows:

$$
M_G : \quad
\begin{array}{c c}
 & \begin{array}{c c c c c} a & b & c & d & e \end{array} \\
\begin{array}{c} a \\ b \\ c \\ d \\ e \end{array} &
\begin{pmatrix}
0 & 1 & 1 & 1 & 1 \\
1 & 0 & 1 & 1 & 0 \\
1 & 1 & 0 & 1 & 0 \\
1 & 1 & 1 & 0 & 1 \\
1 & 0 & 0 & 1 & 0
\end{pmatrix}
\end{array}
\qquad
M_H : \quad
\begin{array}{c c}
 & \begin{array}{c c c c c} v & w & x & y & z \end{array} \\
\begin{array}{c} v \\ w \\ x \\ y \\ z \end{array} &
\begin{pmatrix}
0 & 1 & 0 & 1 & 1 \\
1 & 0 & 1 & 1 & 1 \\
0 & 1 & 0 & 1 & 0 \\
1 & 1 & 1 & 0 & 1 \\
1 & 1 & 0 & 1 & 0
\end{pmatrix}
\end{array}
$$

If we rearrange the vertices in H as w, v, z, y, x, the resulting adjacency matrix M'_H is as follows:

$$
M'_H : \quad
\begin{array}{c c}
 & \begin{array}{c c c c c} w & v & z & y & x \end{array} \\
\begin{array}{c} w \\ v \\ z \\ y \\ x \end{array} &
\begin{pmatrix}
0 & 1 & 1 & 1 & 1 \\
1 & 0 & 1 & 1 & 0 \\
1 & 1 & 0 & 1 & 0 \\
1 & 1 & 1 & 0 & 1 \\
1 & 0 & 0 & 1 & 0
\end{pmatrix}
\end{array}
$$

We can now easily see that $M_G = M'_H$, and thus conclude that G and H are isomorphic.

Thus, if two graphs are isomorphic, an intelligent system of reordering the vertices of both graphs will finally result in two equal adjacency matrices. On the other hand, if the two graphs are not isomorphic, the person, or more likely the computer, will have to exhaust the large number of reorderings to come to a conclusion that no isomorphism exists.

Question 2.2.4.   *Use adjacency matrices to answer Question 2.2.2.*

Exercise 2.2

(1) Draw all non-isomorphic graphs of order n with $1 \leq n \leq 4$.

(2) (i) Draw all non-isomorphic graphs of order 5 and size 3.
  (ii) Draw all non-isomorphic graphs of order 5 and size 7.

(3) Determine if the following two graphs are isomorphic.

(4) Determine if the following two graphs are isomorphic.

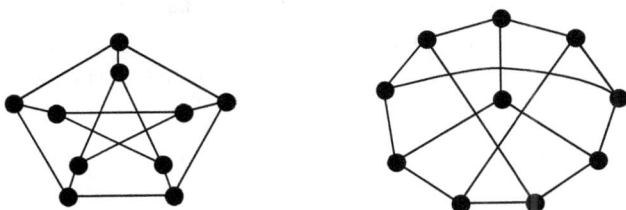

(5) (+) The following two graphs G and H are isomorphic. List all the isomorphisms from G to H.

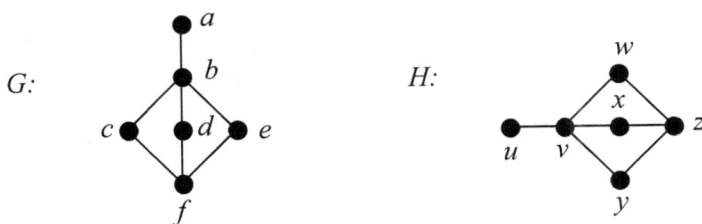

(6) (+) Prove, by definition of an isomorphism, that the relation '$\cong$' is reflexive, symmetric and transitive among the family of graphs; that is, the properties listed in Question 2.1.3.

(7) Let f be an isomorphism from a graph G to a graph H and w a vertex in G. Show that the degree of w in G is equal to the degree of $f(w)$ in H.

(8) (+) A given graph G of order 5 contains at least two vertices of degree 4.

  (i) Assume that not all vertices in G are even. Find all possible degree sequences of G, in non-increasing order; and for each case, construct all such G which are not isomorphic.

  (ii) Assume that all vertices in G are even. Find all possible degree sequences of G, in non-increasing order; and for each case, construct all such G which are not isomorphic.

(9) (+) Let H be a graph of order 5 which contains more odd vertices than even. Find all possible degree sequences of H in non-increasing order; and for each case, construct all such H which are not isomorphic.

(10) Construct two non-isomorphic 3-regular graphs of order 10.

(11) (+) Let G and H be two isomorphic graphs. Show that

  (i) if G is connected, then H is connected;

  (ii) if G is disconnected, then H is disconnected, and they have the same number of components.

(12) Prove that if the adjacency matrices of two graphs G and H are equal, then the graphs G and H are isomorphic.

(13) Using adjacency matrices, determine which, if any, of the following three graphs are isomorphic.

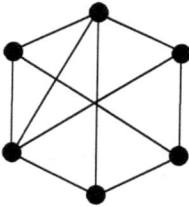

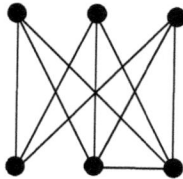

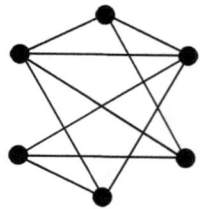

2.3 Subgraphs of a graph

In studying problems on a graph, quite often, we may wish to consider the 'graphical structures' of **certain portions** of the graph. For instance, in determining whether the graphs G and H in Example 2.1.2 are isomorphic, we may try the following way: observe that the vertices x_1, x_2 and x_4 in G form a K_3, but there is no K_3 contained in H; this implies that G and H have different 'graphical structures', and so $G \not\cong H$.

Let G be a graph. A graph H is called a **subgraph** of G if $V(H) \subseteq V(G)$ and $E(H) \subseteq E(G)$.

By definition, every graph is a subgraph of itself.

A subgraph H of G is said to be **proper** if $H \not\cong G$.

Example 2.3.1. *Consider the graphs G, $H_1, H_2, \cdots, H_6$ as shown in Figure 2.11. We observe that*
(1) H_1 is not a subgraph of G as $E(H_1) \not\subseteq E(G)$ though $V(H_1) \subseteq V(G)$;
(2) $H_2, \cdots, H_6$ are subgraphs (indeed, proper subgraphs) of G.

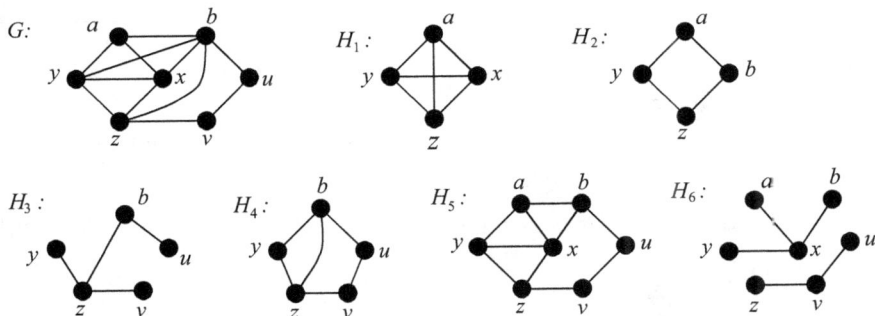

Figure 2.11

Note that $V(H_i) \neq V(G)$ for $i = 2, 3, 4$, but $V(H_5) = V(H_6) = V(G)$.

A subgraph H of a graph G is said to be **spanning** if $V(H) = V(G)$.

Thus, in Example 2.3.1, the graphs H_5 and H_6 are spanning subgraphs of G , but H_2, H_3 and H_4 are not.

Observe that the spanning subgraph H_5 of G can be obtained by deleting the edges by and bz from G. More generally, let F be a set of edges in G. We denote by $G - F$ the subgraph of G obtained by deleting the edges in F from G. Note that whenever an edge uv is deleted, its two ends (namely, u and v) are still in G. If $F = \{e\}$, consisting of a single edge e, we simply write $G - e$ for $G - \{e\}$.

Question 2.3.1.   *Let G be the graph shown in Figure 2.11. Draw the subgraphs $G - yz$, $G - \{uv, vz\}$ and $G - \{yz, bz, ax, uv\}$.*

It can be shown (see Problem 2 of Exercise 2.3) that:

A subgraph H of a graph G is a spanning subgraph of G if and only if H is obtained from G by deleting some edges in G, that is, $H = G - F$ for some $F \subseteq E(G)$.

For instance, in Example 2.3.1, we have $H_5 = G - \{by, bz\}$ and $H_6 = G - \{ay, ab, by, bz, bu, xz, yz\}$.

We shall now show how the notion of **subgraph** is used for isomorphic graphs testing.

Example 2.3.2.   *Consider the graphs G and H as shown in Figure 2.12.*

G: H:

Figure 2.12

Note that G and H have the same degree sequence in non-increasing order, that is, $(3, 3, 2, 2, 2, 2)$, yet $G \ncong H$ (this shows that the converse of Result (2) in Section 2.2 is false). How do we argue that $G \ncong H$?

Some ways using the concept of subgraphs are given below:

(i) G contains one K_3 as a subgraph, but H contains two K_3's;

(ii) the two vertices of degree 3 in G are contained in a common $\bar{K}_3$, but this is not the case in H;

(iii) G contains a spanning subgraph which is a cycle (i.e., C_6), but H does not have one;

(iv) G contains a C_5, but H does not have one;
    etc.

Any one of the reasons above would be good enough to justify that $G \ncong H$.

As we have just seen, to show that $G \ncong H$, all we need is to find a 'property' that G has but H doesn't have (or vice versa). For this purpose, we state below another useful fact in terms of 'subgraphs'. First of all, we introduce the following notation.

For two graphs G and R, let $n_G(R)$ denote the number of subgraphs of G which are isomorphic to R.

Thus, in Example 2.3.2, $n_G(K_3) = 1$ and $n_H(K_3) = 2$, $n_G(C_5) = n_G(C_6) = 1$ and $n_H(C_5) = n_H(C_6) = 0$.

Result (3). Let G and H be graphs such that $G \cong H$. Then for any graph R, $n_G(R) = n_H(R)$.

Equivalently, Result (3) says that if $n_G(R) \neq n_H(R)$ for some graph R, then $G \ncong H$.

Question 2.3.2. *Consider the graphs G and H in Figure 2.13.*

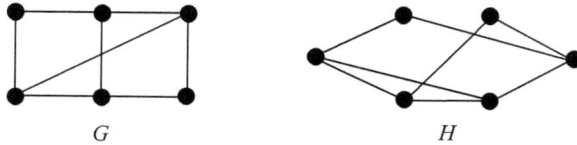

G H

Figure 2.13

Find (i) $n_G(C_3)$ and $n_H(C_3)$,

(ii) $n_G(C_4)$ and $n_H(C_4)$,

(iii) $n_G(C_5)$ and $n_H(C_5)$,

(iv) $n_G(C_6)$ and $n_H(C_6)$.

Is it true that $G \cong H$?

Question 2.3.3. *Let G be a graph of order $n \geq 2$. What is $n_G(K_1)$? What is $n_G(K_2)$? Is Result (1) a special case of Result (3)?*

Look at the subgraphs H_3 and H_4 of G in Example 2.3.1. By comparing these two subgraphs, we notice that while $V(H_3) = \{b, u, v, y, z\} = V(H_4)$, $E(H_3) \neq E(H_4)$. In H_3, some edges in G which join certain pairs of vertices in H_3 are not present; for instance, yb and uv. On the other hand, **every edge** in G which joins a pair of vertices in H_4 always **remains** in H_4. This feature of H_4 motivates the introduction of the following important type of subgraphs of a graph.

> A subgraph H of a graph G is called an **induced subgraph** of G if **any edge** in G that joins a pair of vertices in H is also in H.
>
> If H is an induced subgraph of G, we also say that H is the subgraph **induced by its vertex set** $V(H)$ and we write $H = [V(H)]$.

Thus, in Example 2.3.1, among the subgraphs $H_2, \cdots, H_6$ of G, only H_4 is an induced subgraph of G, and we see that H_4 is induced by $\{b, u, v, y, z\}$ (in notation, $H_4 = [\{b, u, v, y, z\}]$). The subgraphs of G in Example 2.3.1

induced by $\{a, x, y, z\}$ and $\{a, b, u, x, z\}$ are shown in (a) and (b) of Figure 2.14 respectively.

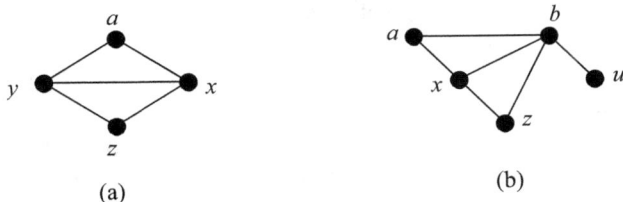

(a)

(b)

Figure 2.14

Question 2.3.4. *Let G be the graph shown in Figure 2.11. Draw the subgraphs of G induced by $\{a, u\}$, $\{a, b, x, y\}$, $\{a, x, y, u, v\}$ and $V(G)$ respectively.*

Question 2.3.5. *Let H be a spanning and induced subgraph of a graph G. What can be said of H?*

Question 2.3.6. *Consider the graph H of Figure 2.15. Note that H is a subgraph of the graph G in Figure 2.11. Is H an induced subgraph of G? Add some more edges in G to H so that the resulting one is $[V(H)]$.*

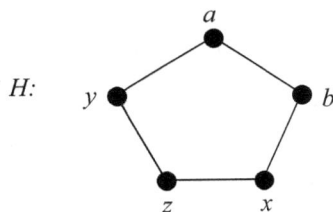

H:

Figure 2.15

Remark. A subgraph H of a graph G is not necessarily an induced subgraph of G. However, it can always be extended to an induced subgraph of G induced by $V(H)$ by adding to H all the missing edges (sharing their two ends in $V(H)$) existing in G.

We have seen that the spanning subgraphs of a graph G are those subgraphs of G that can be obtained from G by deleting some edges in G. In contrast with this, we shall see that induced subgraphs of G can be obtained from G as well, but by deleting some vertices in G as defined below.

Let G be a graph and W a set of vertices in G. We shall denote by $G - W$ the subgraph of G obtained by removing each vertex in W from $V(G)$ together with **all the edges incident with it** from $E(G)$. When W is a singleton, say $W = \{w\}$, we shall write $G - w$ for $G - \{w\}$.

For instance, if G is the graph given in Example 2.3.1, then the subgraphs $G - x$, $G - \{x, y\}$ and $G - \{x, y, z\}$ of G are shown in (a), (b) and (c) of Figure 2.16 respectively. Note that $G - \{x, y, z\} = [\{a, b, u, v\}]$, $G - \{x, y\} = [\{a, b, u, v, z\}]$ and $G - x = [\{a, b, u, v, y, z\}]$.

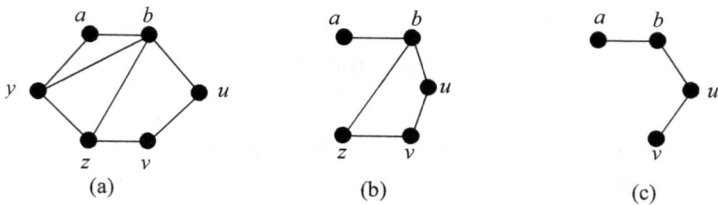

Figure 2.16

In general, one can show (see Problem 4 of Exercise 2.3) that:

> A subgraph W of a graph G is an induced subgraph of G if and only if $W = G - (V(G) \backslash V(W))$, where $V(G) \backslash V(W)$ consists of those vertices of G which are *not* in W.

Question 2.3.7. *We have introduced various concepts of subgraphs of a graph. Could they be generalized to multigraphs?*

To end this section, we introduce the following:

The Reconstruction Conjecture

Let us begin with a simple problem. Given a graph G with four vertices v_1, v_2, v_3 and v_4 together with the following information:

(i) $G - v_1 \cong$

(ii) $G - v_2 \cong$

(iii) $G - v_3 \cong$

(iv) $G - v_4 \cong$

What is G?

First of all, by (i), G contains the following graph as a subgraph:

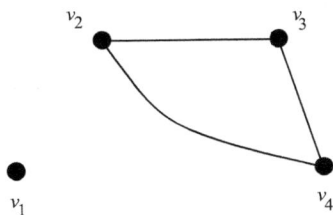

Then, by (iii), G must contain the graph of Figure 2.17 as a subgraph.

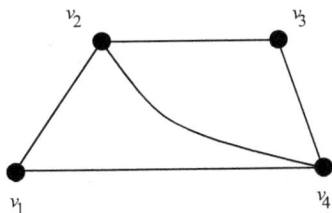

Figure 2.17

Now, it is easily seen that this graph fulfills (ii) and (iv). Furthermore, it can be checked that this graph is the **only** graph that fulfills (i) to (iv). We thus conclude that G is the graph of Figure 2.17.

Let G be a graph with $V(G) = \{u_1, u_2, \cdots, u_n\}$. We say that G is **reconstructible** if, whenever H is a graph with $V(H) = \{v_1, v_2, \cdots, v_n\}$ such that $H - v_i = G - u_i$ for each $i = 1, 2, \cdots, n$, then $H \cong G$ (that is, G is uniquely determined by its n subgraphs: $G - u_1, G - u_2, \cdots, G - u_n$).

Thus, the above example shows that the graph of Figure 2.17 is reconstructible. A very well-known unsolved problem in graph theory can now be stated below.

The Reconstuction Conjecture. Every graph of order at least three is reconstructible.

We note that a graph of order two is not reconstructible. Indeed, take G and H as shown below:

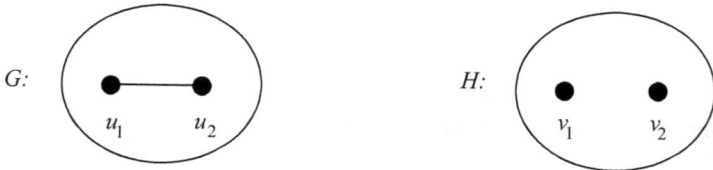

It is observed that $G - u_1 \cong H - v_1$ and $G - u_2 \cong H - v_2$, and yet $G \not\cong H$.

The above conjecture was first posed by a famous scientist S. M. Ulam (see also [U]) and was initially studied by P. J. Kelly in his Ph.D. thesis around 1942 (see [Ke]). Though the conjecture has been verified to be true for some special families of graphs such as regular graphs and disconnected graphs, it remains unsettled for the general situation. For a very general survey on this conjecture, the reader is referred to the excellent article [B] by Bondy.

Exercise 2.3

(1) Let G be the graph given below:

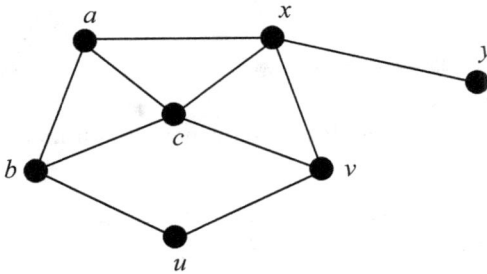

(i) Draw the subgraphs $[\{b, v, y\}]$, $[\{a, b, c, v, x\}]$ and $[\{a, b, u, v, x\}]$ of G.
(ii) Draw the subgraphs $G - \{ab, cv, xy\}$, $G - c$ and $G - \{b, v\}$ of G.
(iii) Find $E([\{a, b, c, x\}])$.
(iv) Draw the subgraph $G - E([\{a, b, c, x\}])$ of G.
(v) Draw a spanning subgraph of G that is connected and that contains a unique C_3 as a subgraph.
(vi) Draw a spanning subgraph of G that is connected and that contains no cycle as a subgraph.

(2) Let H be a subgraph of a graph G. Show that H is a spanning subgraph of G if and only if $H = G - F$, where $F \subseteq E(G)$.

(3) Let G be a graph and $X \subseteq V(G)$. Show that $G - X = [V(G)\backslash X]$.

(4) Let G be a graph and W a subgraph of G. Show that W is an induced subgraph of G if and only if $W = G - (V(G)\backslash V(W))$.

(5) Determine which of the following four graphs are isomorphic and which are not so.

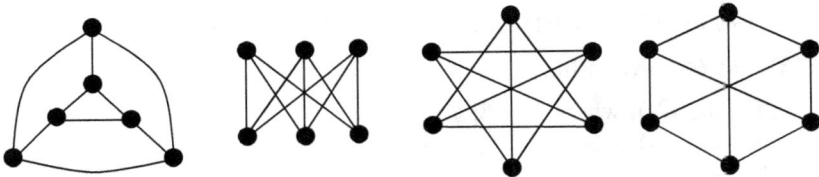

(6) Let G and H be the two graphs given below:

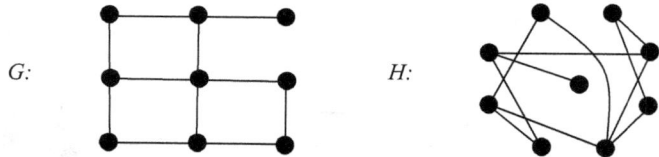

G:

H:

Do they have the same degree sequence in non-increasing order? Are they isomorphic?

(7) (+) Let G be a graph of order five satisfying the following condition: for any three vertices x, y, z in G, $[\{x, y, z\}]$ is not isomorphic to either N_3 or K_3.
What is the graph G? Justify your answer.

(8) Draw all non-isomorphic graphs of order 5 which contain a C_5.

(9) Let H be a spanning subgraph of a graph G. Which of the following statements is/are true? Why?

  (i) If G is connected, then H is connected.
 (ii) If H is connected, then G is connected.

(10) Let G be a disconnected graph with k components. Choose a vertex from each component. What is the subgraph induced by these k vertices?

(11) For a graph G, denote by $c(G)$ the number of components in G. Thus, for the graph G below, $c(G) = 4$.

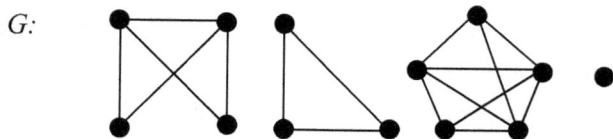

G:

Let H be a spanning subgraph of a graph G. Show that $c(H) \geq c(G)$.

(12) (+) Let C be a cycle and S a subset of $V(C)$. Show that $c(C-S) \leq |S|$.

(13) Let $G = K_n$. Find

  (i) $n_G(C_3)$, where $n \geq 3$;
 (ii) $n_G(C_4)$, where $n \geq 4$; and
(iii) $n_G(C_k)$, where $n \geq k \geq 5$.

(14) Let G be the Petersen graph. Find $n_G(C_i)$, where $i = 3, 4, 5$. What is the largest cycle in G?

(15) Let G be a graph of order 5 which contains at least two vertices of degree 4 and a C_5. Find all possible degree sequences of G, in non-increasing order; and for each case, construct all such G.

(16) Let G be a connected graph. An edge e in G is called a **bridge** if $G - e$ is disconnected.

    (i) Find all bridges in the following graph:

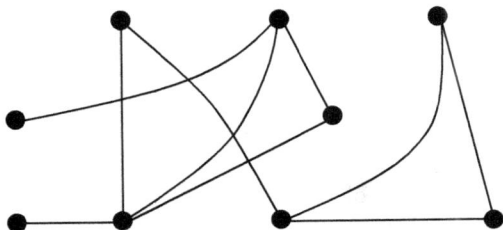

    (ii) How many components does $G - e$ have if e is a bridge in G?

    (iii) Show that an edge e in G is a bridge if and only if e is not contained in any cycle in G.

(17) (+) Let G be a connected graph in which every vertex is even. Show that G contains no bridges.

(18) Let G be a connected graph. A vertex w in G is called a **cut-vertex** if $G - w$ is disconnected.

    (i) Find all cut-vertices in the following graph:

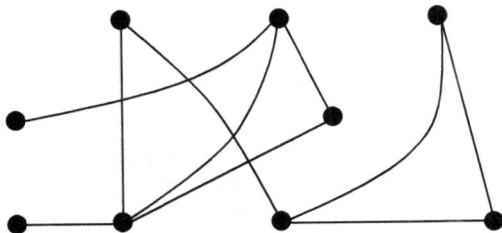

    (ii) How many components does $G - w$ have if w is a cut-vertex of G?

    (iii) (+) Assume that $v(G) \geq 3$. Show that if G contains a bridge, then G contains a cut-vertex.

    (iv) Is the converse of (iii) true?

(19) (∗) Let G be a cubic (i.e., 3-regular) graph. Suppose G contains a cut-vertex. Must G contain a bridge? Why?

(20) Let G be a connected graph of order 8 and size 12 which contains no bridges. Suppose that $\Delta(G) = 4$ and G has exactly two vertices of degree 4.

   (i) Find the number of end-vertices in G.

   (ii) Find the number of vertices of degree 3 in G.

   (iii) Construct three such graphs which are non-isomorphic.

(21) (+) Show that a graph G contains a cycle of length at least $\delta(G) + 1$ if $\delta(G) \geq 2$.

(22) Let G be a graph of order 9. Assume that $\Delta(G) = 6$ and that G contains at least 4 vertices of degree at least 4. Show that G contains a C_3.

(23) Let G be a graph of order n with degree sequence $(d_1, d_2, \cdots, d_n)$. Construct a graph from G having the degree sequence $(d_1 + 1, d_2 + 1, \cdots, d_n + 1, n)$.

(24) (+) Let G be a connected graph of order n. Show that the vertices in G can always be named as $x_1, x_2, \cdots, x_n$ such that the induced subgraph $[\{x_1, x_2, \cdots, x_i\}]$ is connected for each $i = 1, 2, \cdots, n$.

(25) Let G be a connected graph of order 8 which contains two C_4's having no vertex in common.

   (i) What is the least possible value of $e(G)$?

   (ii) Assume that G contains no cut-vertices. What is the least possible value of $e(G)$?

   (iii) Assume that G contains no odd vertices. What is the least possible value of $e(G)$?

   (iv) Assume that G contains no even vertices. What is the least possible value of $e(G)$?

   For each of the above cases, construct a corresponding G which has its $e(G)$ attaining your least possible value.

(26) Let G be a graph with $V(G) = \{x_1, x_2, x_3, x_4\}$ such that $G - x_1 \cong$ •⌒⌒⌒• , $G - x_2 \cong$ • •——• , $G - x_3 \cong$ •——•——• and $G - x_4 \cong$ •——•——• .

Determine G and justify your answer.

(27) (+) Let G be a graph with $V(G) = \{y_1, y_2, \cdots, y_5\}$ such that

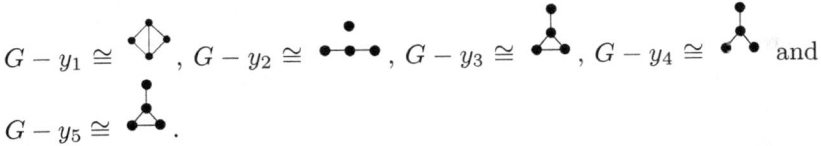

$G - y_1 \cong$, $G - y_2 \cong$, $G - y_3 \cong$, $G - y_4 \cong$ and

$G - y_5 \cong$.

Determine G and justify your answer.

(28) (+) Let G be a graph with $V(G) = \{u_1, u_2, \cdots, u_n\}$, where $n \geq 3$. Let $m = e(G)$, $m_i = e(G - u_i)$, $i = 1, 2, \cdots, n$. Show that

  (i) the degree of u_i in G is equal to $m - m_i$, $i = 1, 2, \cdots, n$;

  (ii) $m = (m_1 + m_2 + \cdots + m_n)/(n - 2)$.

(29) (+) Let G be a connected multigraph of order at least two and A be a subset of $V(G)$. Denote by $e(A, V(G) \backslash A)$ the number of edges having one end in A and the other in $V(G) \backslash A$.

  (i) Let H be the multigraph shown below and $A = \{u, v, z\}$. Find $e(A, V(H) \backslash A)$.

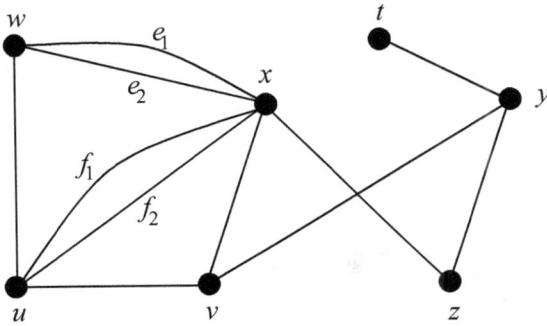

  (ii) Show that $e(A, V(G) \backslash A)$ is even if and only if A contains an even number of odd vertices in G.

2.4 The complement of a graph

The graph of Figure 2.18 shows the 'acquaintance' relationship among a group of five people: a, b, c, d and e. The five people are represented by five vertices, and two vertices are adjacent if and only if they are mutual acquaintances (assuming that this relation is symmetric). Thus, a and b are acquaintances, so are a and e, b and e, e and c, and c and d.

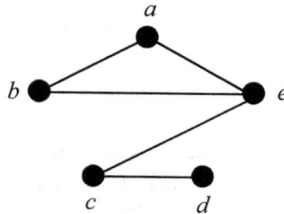

Figure 2.18

Based on the graph of Figure 2.18, we construct a new graph on the same vertex set in Figure 2.19 which now shows the 'stranger' relationship among the five.

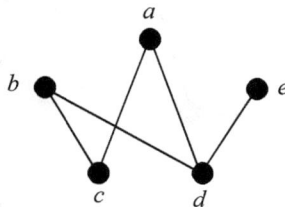

5Figure 2.19

Figure 2.19

What is the relation between these two as graphs?

Firstly, they have the **same vertex set**. Secondly, if two vertices are **adjacent** in the first graph (such as a and b), then they are **not adjacent** in the second; and if two vertices are **not adjacent** in the first (such as a

and c), then they are **adjacent** in the second. We call the second graph the **complement** of the first, and vice versa.

For a given graph G, the **complement** of G, denoted by $\overline{G}$, is the graph with $V(\overline{G}) = V(G)$ such that two vertices are adjacent in $\overline{G}$ if and only if they are not adjacent in G.

Thus $v(G) = v(\overline{G})$. If $v(G) = 1$, then note that $\overline{G} \cong G \cong K_1$.

Example 2.4.1.   *Figure 2.20 displays six graphs G of order 5. The complements of the first three are shown. You are invited to construct the complements of the remaining three.*

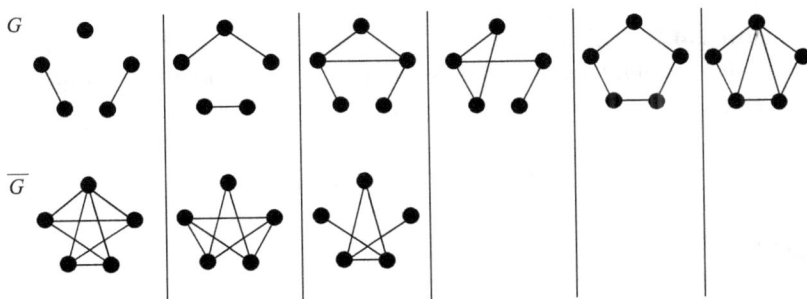

Figure 2.20

Question 2.4.1.   (1) *What is $\overline{N_n}$?*

(2) *What is $\overline{K_n}$?*

(3) *Define $\overline{\overline{G}}$ as $\overline{(\overline{G})}$. What is $\overline{\overline{G}}$?*

(4) *Suppose G and H are two graphs such that $G \cong H$. Is it true that $\overline{G} \cong \overline{H}$? (See Problem 3 of Exercise 2.4.)*

(5) *If G is a graph with degree sequence $(4, 4, 3, 3, 2, 2)$, what is the degree sequence of $\overline{G}$, in non-increasing order?*

(6) *Let G be a 3-regular graph of order 10. Is $\overline{G}$ also regular? If 'yes', what is the degree of each vertex in $\overline{G}$?*

(7) *In Example 2.4.1, for each pair G and $\overline{G}$, if we superimpose $\overline{G}$ onto G so that the same vertices are identified, what is the resulting graph? Is it a K_5?*

(8) *What is the sum $e(G) + e(\overline{G})$ for each of the six graphs G in Example 2.4.1?*

In general, given a graph G of order n, by superimposing $\overline{G}$ onto G so that the same vertices are identified, we would obtain the complete graph K_n. Thus, we have:

Result (4). For any graph G of order n, $e(G) + e(\overline{G}) = e(K_n) = \binom{n}{2}$.

Question 2.4.2.   *In Figure 2.20, we notice that the first graph is disconnected but its complement becomes connected. Consider the disconnected graph H of Figure 2.21.*

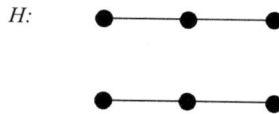

Figure 2.21

Construct $\overline{H}$. Is $\overline{H}$ connected?

Yes! The above $\overline{H}$ is also connected. Indeed, we have the following general result:

Result (5).   Let G be a graph. If G is disconnected, then $\overline{G}$ is connected.

Proof.   Assume that G is disconnected. To show that $\overline{G}$ is connected, we show that every two vertices in $\overline{G}$ are joined by a path in $\overline{G}$. Thus, let

$u, v \in V(\overline{G}) (= V(G))$. If u, v are in different components in G, then u and v are joined by an edge in $\overline{G}$. If u and v are in the same component of G, let w be any vertex in another component of G, then uwv is a $u - v$ path in $\overline{G}$ (see Figure 2.22). This completes the proof. □

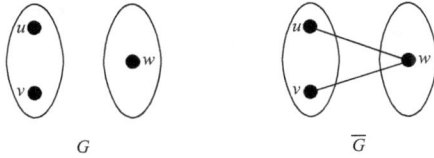

Figure 2.22

Question 2.4.3. *For a disconnected graph G, the above proof not just shows that $\overline{G}$ is connected, but further reveals that the distance between any two vertices is very small in $\overline{G}$. How small is it?*

Question 2.4.4. *Which of the graphs G in Figure 2.20 are such that $G \cong \overline{G}$?*

We are sure that you could get the right answer for Question 2.4.4, namely, the third and fifth graphs. Graphs of this type are interesting and have attracted some researchers' attention.

A graph G is said to be **self-complementary** if $G \cong \overline{G}$.

Thus, the third and fifth graphs in Figure 2.20 are self-complementary.

Question 2.4.5. *Construct a self-complementary graph of order n, where $2 \le n \le 4$.*

Self-complementary graphs are quite rare. Indeed, among all graphs of order n, $2 \le n \le 7$, there are only three such graphs. Also, self-complementary graphs of order n are available only for some types of n as stated below (see Problem 13 of Exercise 2.4)

Result (6). Let G be a self-complementary graph. Then
(i) G is connected and
(ii) $v(G) = 4k$ or $v(G) = 4k + 1$ for some integer k.

Exercise 2.4

(1) Consider Problem 2 of Exercise 2.2. Is there any relation between the family of graphs found in (i) and the family of graphs in (ii)?

(2) (i) Draw all non-isomorphic graphs of order 6 and size 3.
    (ii) Find the number of non-isomorphic graphs of order 6 and size 12.

(3) Let G and H be two graphs. Show that $G \cong H$ if and only if $\overline{G} \cong \overline{H}$.

(4) Let G be a disconnected graph. Show that the distance between any two vertices in $\overline{G}$ is at most two. (See the proof of Result (5).)

(5) Let G be a k-regular graph of order n. Is $\overline{G}$ also regular? If the answer is 'yes', what is the degree of each vertex in $\overline{G}$?

(6) Let G be a graph of order n with degree sequence $(d_1, d_2, \cdots, d_n)$ in non-increasing order. Find the degree sequence of $\overline{G}$ in non-increasing order.

(7) Draw all non-isomorphic 4-regular graphs of order 7.

(8) How many non-isomorphic graphs are there with degree sequence $(5, 5, 4, 4, 4, 4)$? Construct one such graph.

(9) How many non-isomorphic graphs are there with degree sequence $(5, 5, 5, 4, 4, 3)$? Construct one such graph.

(10) (+) What is the value of each diagonal entry in the matrix $A(G)A(\overline{G})$?

(11) For each of the following graphs,

   (i) construct its complement and
   (ii) determine if it is self-complementary.

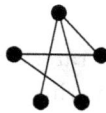
(a)

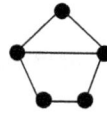
(b)

(c)

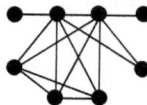
(d)

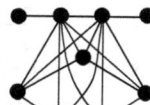
(e)

(12) (+) Show that every self-complementary graph is connected.

(13) (+) Let G be a self-complementary graph of order $n \geq 2$. Show that
   (i) $e(G) = \frac{1}{4}n(n-1)$, and
   (ii) $n = 4k$ or $n = 4k+1$ for some positive integer k.

(14) Determine the values of n, where $n \geq 3$, for which the cycle C_n is self-complementary.

(15) Let G be a self-complementary graph of order n. Show that if G is regular, then $n = 4k+1$ for some positive integer k.

(16) Construct a regular self-complementary graph of order 9.

(17) (+) (i) Let G be a self-complementary graph of order 9. Show that G contains at least one vertex of degree 4.
   (ii) Generalize the result in (i).

(18) Let G be a graph and x be a vertex in G.
   (i) Is it true that $\overline{G - x} = \overline{G} - x$?
   (ii) If x is a cut-vertex of G, is $\overline{G} - x$ connected?

   Justify your answers.

(19) (+) (i) Show that at a gathering of any six people, some three of them are either mutual acquaintances or complete strangers to one another.
   (ii) Does the result in (i) still hold for 'five' people?

(20) (+) Let G be a graph of order 6. If G does not contain N_3 as an induced subgraph, what is the least possible value for $n_G(C_3)$?

(21) Let G be a graph with $\Delta(G) \geq r$, where r is a positive integer. Show that either G contains a triangle or $\overline{G}$ contains a K_r.

(22) (+) Let G be a graph of odd order and $\delta(G) \geq 5$. Assume that G contains no N_3 as an induced subgraph. Show that G contains a K_4.

(23) (+) Let G be a graph of order n which contains no triangles.
   (i) Assume that $n = 9$. Show that $\overline{G}$ contains a K_4.
   (ii) Assume that $n = 8$. Must $\overline{G}$ contain a K_4?

(24) (+) Seventeen people correspond by mail with one another – each one with all the rest. In their letters only three different topics are discussed. Each pair of correspondents deals with only one of these topics. Prove that there are at least three people who write to each other about the same topic. (IMO 1964/4)

Chapter 3

Bipartite Graphs and Trees

3.1 Bipartite graphs

In Example 1.2.3, the following graph which models a job-application situation is shown:

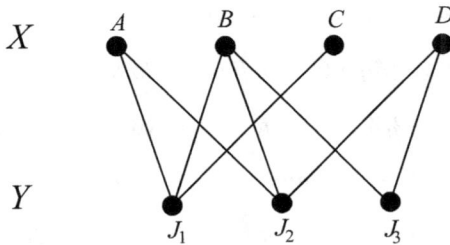

This graph is of order 7 and size 8, and the above diagram is drawn in such a way that the seven vertices are divided into two parts, X and Y, such that each of the eight edges has an end in X and the other in Y.

Another example of this type of graphs is shown below:

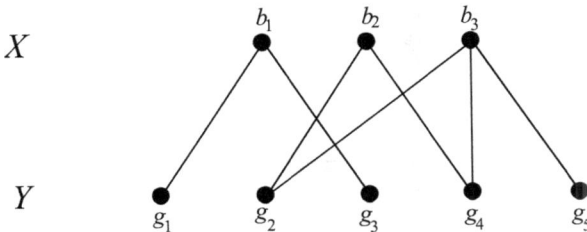

Here X is a group of 3 boys and Y is a group of 5 girls in a party, and that 'x' and 'y' are adjacent means that the boy 'x' dances with the girl 'y' in the party. Graphs of this type are very important and useful. We call them bipartite graphs.

> A graph G of order at least 2 is said to be **bipartite** if its vertex set $V(G)$ can be partitioned into two non-empty subsets X and Y such that each of the edges in G joins a vertex in X to a vertex in Y. We call X and Y the **partite sets**, and call the pair (X, Y) a **bipartition** of G.

Remark. The graph K_1 is considered as a special bipartite graph.

Question 3.1.1. *A bipartite graph G with a bipartition (X, Y) is defined as follows:* $X = \{a, b, c, d\}$ *and* $Y = \{u, v, w\}$, *and* $E(G) = \{aw, bu, bw, cv, du, dv, dw\}$. *Draw the graph G.*

Question 3.1.2. *A graph G is defined as follows:* $V(G) = X \cup Y$, *where*

$$X = \{2, 3, 5\} \quad and \quad Y = \{6, 7, 8, 9, 10\},$$

and $E(G) = \{xy | x \in X, y \in Y$ *and y is divisible by $x\}$.*

(i) Is '2' adjacent to '6'? Is '3' adjacent to '8'? Is '5' adjacent to '10'?

(ii) Draw the graph G.

(iii) Is G bipartite?

Consider the graph H of Figure 3.1. Is it a bipartite graph?

H:

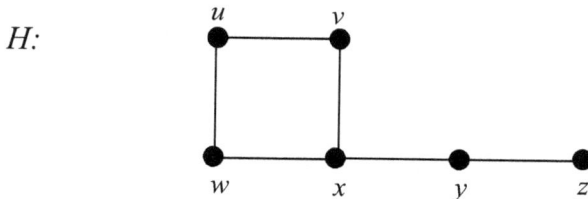

Figure 3.1

To answer this question, by definition, we ask ourselves: Does H have a bipartition (X, Y)?

If we take $A = \{u, v\}$ and $B = \{w, x, y, z\}$, is (A, B) a bipartition of H? The answer is 'no'. Why? Since there is an edge joining two vertices in A.

Though, in this case, (A, B) is not a bipartition of H, it does not mean that H is not bipartite. Indeed, if we take $X = \{u, x, z\}$ and $Y = \{w, v, y\}$, then we observe that

(i) $V(H) = X \cup Y$ and

(ii) each edge in H joins a vertex in X to a vertex in Y.

It follows by definition that (X, Y) is a bipartition of H, and thus H is a bipartite graph.

Let us re-draw the graph H in a 'natural' way as shown in Figure 3.2. It is now clear that H is a bipartite graph.

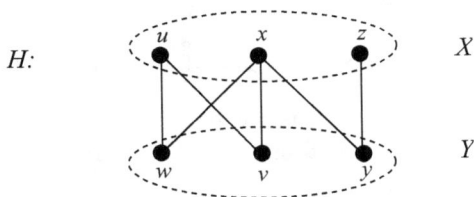

Figure 3.2

Quite often, bipartite graphs are not drawn in this 'natural' form. How to find a *bipartition* of a bipartite graph thus becomes a practical and interesting problem. We shall discuss this problem later.

Question 3.1.3. *Determine whether each of the graphs in Figure 3.3 is bipartite. If it is so, find a corresponding bipartition.*

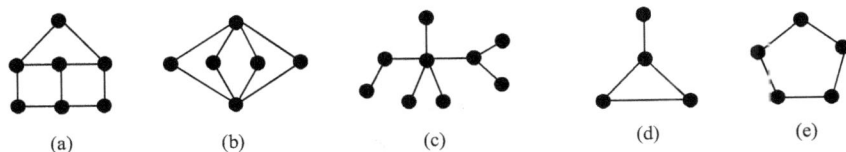

Figure 3.3

In Figure 3.3, while the graphs (a), (b) and (c) are bipartite, the graphs (d) and (e) are not. Let us examine, for instance, the graph (e), which is C_5. For convenience, we name its five vertices as shown in the following:

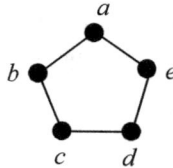

Suppose on the contrary that it is bipartite and has its bipartition (X, Y). We may assume that $a \in X$. As no two adjacent vertices can be in the same part, $b \in Y$. This, in turn, implies (anti-clockwise) that $c \in X$, $d \in Y$ and $e \in X$. We thus arrive at the situation that both a and e are in X and they are adjacent, which, however, is not allowed. This shows that C_5 has no bipartition, and is therefore not bipartite.

A cycle C_k is said to be **odd** if k is odd, and **even** if k is even.

From the above discussion, it is not hard to see that the argument can be similarly carried out to lead to a contradiction if the graph contains an odd cycle. Thus, we conclude that

       if G contains an odd cycle, then G is not bipartite;

or equivalently,

       if G is bipartite, then G contains no odd cycles.

Is the converse of this result true? That is, *if G contains no odd cycles, must G be bipartite?*

Question 3.1.4. *The graphs (a), (b) and (c) in Figure 3.3 contain no odd cycles and they are bipartite. Consider the graph of Figure 3.4.*

    (i) Does it contain any odd cycle?

    (ii) Is it bipartite?

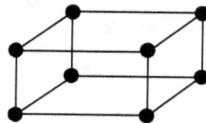

Figure 3.4

Yes! If a graph contains no odd cycles, then it must be bipartite! This result was found by the Hungarian combinatorist, Denes König (1884-1944), in 1916. König wrote the first book [Kö] on graph theory in 1936.

Theorem 3.1. *A graph G is bipartite if and only if it contains no odd cycles.*

The proof of the 'necessity', namely, 'if G is bipartite, then G contains no odd cycles', can be carried out as how we did above, and is left to the reader (see Problem 3 of Exercise 3.1).

We shall now prove the 'sufficiency'; namely, 'if G contains no odd cycles, then G is bipartite'.

For this purpose, we first prove the following simple observation:

Lemma 3.2. *Every closed walk of odd length in a graph always contains an odd cycle.*

Proof. Let W be a closed walk of odd length $p \geq 3$. We shall prove the statement by induction on p.

For $p = 3$, we have $W = w_1 w_2 w_3 w_1$, which obviously forms a C_3. Assume that it is true for all closed walks of odd length less than p, where $p \geq 5$.

Now consider a closed walk of odd length $p : W = w_1 w_2 \cdots w_p w_1$. Our aim is to show that W contains an odd cycle. If W forms itself a C_p, we are through; otherwise, some vertices are repeated and there exist i and j with $1 \leq i < j < p$ such that $w_i = w_j$, as shown in the next page.

Consider the two closed walks of smaller length:

$$W_1 = w_1 w_2 \cdots w_i w_{j+1} \cdots w_p w_1$$
$$\text{and } W_2 = w_i w_{i+1} \cdots w_{j-1} w_j \text{ (note that } w_i = w_j\text{)}.$$

One of them must be of odd length (why?), say W_1. As the (odd) length of W_1 is less than p, by the induction hypothesis, W_1 contains an odd cycle. Clearly, this odd cycle is contained in W. The proof is thus complete. $\square$

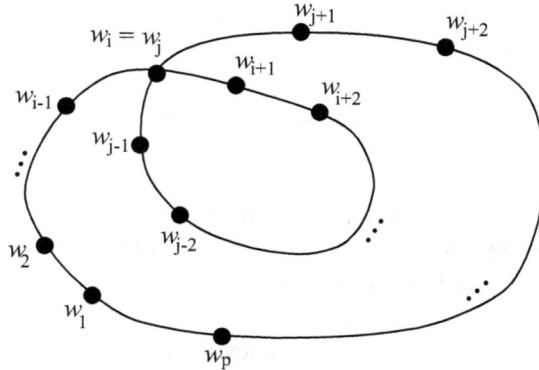

With the help of Lemma 3.2, we are now ready to prove the sufficiency of Theorem 3.1.

Proof. Let G be a graph that does not contain any odd cycles. We aim to show that G is bipartite by providing a bipartition of G.

We may assume that G is connected (why?). Let w be a fixed vertex in G and let

$$X = \{v \in V(G) | d(w,v) \text{ is even}\}$$

and $\quad Y = \{v \in V(G) | d(w,v) \text{ is odd}\}.$

We now claim that (X, Y) **is a bipartition of** G.

It is obvious that X and Y are disjoint, and as G is connected, $V(G) = X \cup Y$ (note that $w \in X$). It remains to show that each edge in G joins a vertex in X to a vertex in Y. Suppose this is not the case. Then there exist u, v in X or u, v in Y, say the former, such that $uv \in E(G)$. As u, v are in X, by definition, $d(w, u)$ and $d(w, v)$ are even, and there exist a $w - u$ path P of even length and a $v - w$ path Q of even length in G.

Consider the walk W which begins at w, follows P to reach u, passes 'uv' to reach v, and finally follows Q to return to w. This walk W is closed and of odd length (why?). By Lemma 3.2, W contains an odd cycle. But then this implies that G contains an odd cycle, a contradiction.

The proof is thus complete.   □

Remark. The argument given in the proof above suggests a way to find a bipartition of a connected graph G if G contains no odd cycles. The procedure is as follows:

Begin with an (arbitrary) vertex and label it '1'. Suppose a vertex has been labeled '1', label all its neighbours '2'; and if a vertex has been labeled '2', label all its neighbours '1'. Repeat this until all vertices have been labeled.

Then the set of all vertices with label '1' and the set of vertices with label '2' form a bipartition of G.

Question 3.1.5. *Apply the above procedure to find a bipartition for each of the graphs in Figure 3.1, Figure 3.3(a), (b) and (c), and Figure 3.4.*

Question 3.1.6. *The graph of Figure 3.5 is not bipartite. Apply the above procedure to identify an odd cycle.*

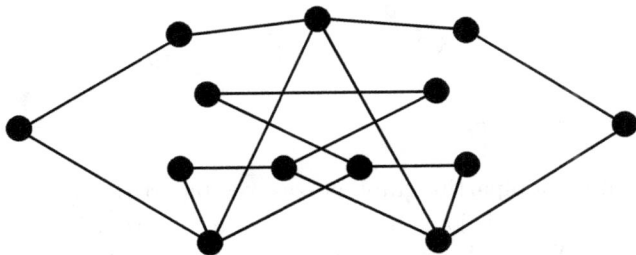

Figure 3.5

Example 3.1.1. *Problem 7 of Exercise 2.3 is restated below:*

Let G be a graph of order five satisfying the following condition: for any three vertices x, y, z in G, $[\{x, y, z\}]$ is not isomorphic to either N_3 or K_3. What is the graph G? Justify your answer.

There are different methods in solving this problem. Here, we solve it by applying Theorem 3.1.

Suppose G is bipartite with a bipartition (X, Y). As $v(G) = 5$, either $|X| \geq 3$ or $|Y| \geq 3$. In either case, G contains an N_3 as an induced subgraph, a contradiction.

Thus, G is non-bipartite. Now, by Theorem 3.1, G contains an odd cycle C_k. By assumption, $k \neq 3$. As $v(G) = 5$, $k = 5$ and G contains C_5 as a spanning subgraph. If $G \neq C_5$, then G would contain a K_3, a contradiction. We therefore conclude that $G = C_5$.

The Handshaking Lemma states that for any multigraph G,
$$\sum_{v \in V(G)} d(v) = 2e(G).$$
Suppose now that G is a bipartite graph with a bipartition (X, Y). Then $\sum_{v \in V(G)} d(v)$ can be split naturally into two parts, namely, $\sum_{x \in X} d(x)$ and $\sum_{y \in Y} d(y)$. What can we say about these two sums? Indeed, it is easily seen that each of them counts $e(G)$. This simple but useful relation is stated below.

Result (1). Let G be a bipartite graph with a bipartition (X, Y). Then

$$\sum_{x \in X} d(x) = e(G) = \sum_{y \in Y} d(y).$$

By definition, a bipartite graph possesses a bipartition (X, Y) such that each edge in the graph joins a vertex in X to a vertex in Y. Note that the null graph N_n is a bipartite graph. Note also that we *do not require* that *every vertex in X must be adjacent to every vertex in Y*. Indeed, this extreme case happens only in a special family of bipartite graphs.

A bipartite graph with a bipartition (X, Y) is called a **complete bipartite graph** if each vertex in X is adjacent to each vertex in Y.

Question 3.1.7. *Which of the following bipartite graphs are complete bipartite graphs?*

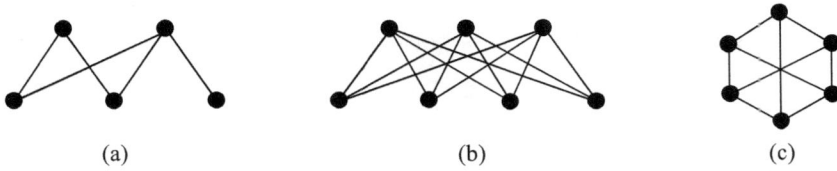

(a)   (b)   (c)

There is one and only one (up to isomorphism) *complete* bipartite graph with a given bipartition (X, Y). If $|X| = p$ and $|Y| = q$, we shall denote this complete bipartite graph by $K(p, q)$ or $K_{p,q}$. In particular, we call $K(1, q)$ or $K(p, 1)$ a **star** (see Figure 3.6).

$K(1,5)$:

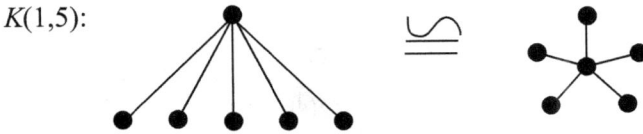

Figure 3.6

Question 3.1.8.

(i) Draw $K(3,5)$ and $K(5,3)$.

(ii) Is $K(3,5)$ isomorphic to $K(5,3)$?

(iii) Find $e(K(3,5))$.

(iv) Find the degree of each vertex in $K(3,5)$.

Remark. For all positive integers p and q, we have:

(1) $K(p, q) \cong K(q, p)$;

(2) $e(K(p, q)) = pq$; and

(3) in $K(p, q)$, the vertices in X are of degree q while those in Y are of degree p.

Exercise 3.1

(1) For each of the following cases, construct all desired connected bipartite
    graphs H of order n:
   (i) $2 \leq n \leq 4$;
   (ii) $n = 5$ and H contains no cycles;
   (iii) $n = 5$ and H contains a cycle;
   (iv) $n = 6$ and H contains a C_6;
   (v) $n = 8$, H is 3-regular and contains a C_8.

(2) Let G be a connected bipartite graph. Then G has a bipartition (X, Y).
    Is $\{X, Y\}$ always unique? What if G is disconnected?

(3) Show that if G is bipartite, then G contains no odd cycles.

(4) Let G be a bipartite graph with a bipartition (X, Y). Show that if G
    is k-regular, where $k \geq 1$, then $|X| = |Y|$.

(5) Construct all non-isomorphic graphs of order 8 and size 10 that are
    bipartite and contain a C_8.

(6) Let G be a bipartite graph of order n with a bipartition (X, Y). Assume
    that G contains a cycle C_n. What is the relation between $|X|$ and $|Y|$?

(7) Does there exist a bipartite graph with degree sequence
    $(5, 5, 5, 4, 4, 3, 3, 3, 1, 1, 1, 1)$?
    Justify your answer.

(8) (+) Show that there does not exist a bipartite graph with degree se-
    quence $(6, 6, \cdots, 6, 5, 3, 3, \cdots, 3)$.

(9) (+) At a party, assume that no boy dances with every girl but each girl
    dances with at least one boy. Prove that there are two couples b, g and
    b', g' which dance, whereas b does not dance with g' nor does g dance
    with b'. (Putnam Exam (1965))

(10) (+) Let G be a bipartite graph with a bipartition (X, Y). Assume that
    $e(G) = v(G)$, and that $d(x) \leq 5$ for each x in X. Show that $|Y| \leq 4|X|$.

(11) (+) Let H be a bipartite graph with a bipartition (X, Y). Assume that
    $e(H) \leq 2v(H)$, and $d(x) \geq 3$ for each x in X. Show that $|X| \leq 2|Y|$.
    Construct one such graph H with $|X| = 2|Y|$.

(12) Let G be a bipartite graph of order $2k$, where k is a positive integer.
    What is the maximum size of G? Find all such bipartite graphs with
    maximum size.

(13) Let G be a bipartite graph of order $2k+1$, where k is a positive integer. What is the maximum size of G? Find all such bipartite graphs with maximum size.

(14) Find, in terms of p and q, the number of C_4 in $K(p,q)$, where $2 \le p \le q$.

(15) Find, in terms of p and q, the number of C_6 in $K(p,q)$, where $3 \le p \le q$.

(16) Let H be a graph obtained from $K(p,q)$, $2 \le p \le q$, by adding a new edge.

   (i) Is H bipartite?
   (ii) What is the largest number of triangles that H could contain?
   (iii) (+) What is the largest number of C_5 that H could contain?

(17) What is the largest cycle in $K(p,q)$, where $2 \le p \le q$?

(18) What can be said about the complement of $K(p,q)$?

(19) Let G be a bipartite graph.

   (i) Is $\overline{G}$ also bipartite?
   (ii) Is $\overline{G}$ always connected?
   (iii) What conditions should be imposed on G so that $\overline{G}$ is connected?

(20) (+) A connected graph G has the following property:
   for each pair of distinct vertices u and v, either all $u - v$ paths are of even length or all $u - v$ paths are of odd length.
   What can be said about G? Justify your answer.

(21) (+) Let G be a graph. A cycle C in G is said to be **induced** if C is induced by $V(C)$.

   (i) Consider the following graph H. Which cycles in H are induced cycles?

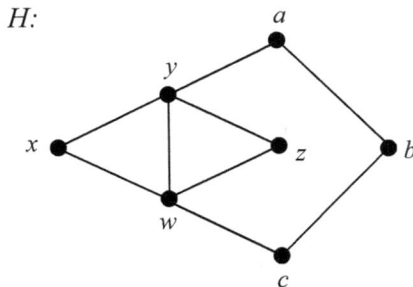

(ii) Show that G is bipartite if and only if G contains no *induced cycles* of *odd* order.

(22) (+) A graph H has the property that each edge in H is incident with an even vertex and an odd vertex. What can be said about H? Construct one such H.

(23) Let G be a bipartite graph of order 7 such that every vertex in G is contained in a cycle.

    (i) Construct one such G.

    (ii) Must G be connected?

    (iii) What is the least possible value of $e(G)$?

    (iv) Construct all non-isomorphic graphs G which have their $e(G)$ attaining the least possible value obtained in (iii).

(24) Let G be a connected bipartite graph of order $p+q$ and size pq, where $1 \le p \le q$. Is it true that $G \cong K(p,q)$?

(25) (+) Let G be a bipartite graph of order $p+q$ and size pq, where $2 \le p \le q$, and with $\delta(G) \ge 1$. Show that $G \cong K(p,q)$ if and only if every two edges in G are contained in a common C_4.

3.2 Trees

In Figure 3.3(c), we show a special bipartite graph, which is *connected* and *contains no cycles*. It is one of a very important family of graphs that we shall study in this section.

A graph is called a **tree** if it is connected and contains no cycles.

Question 3.2.1. *Which of the graphs in Figure 3.7 is a tree? Why?*

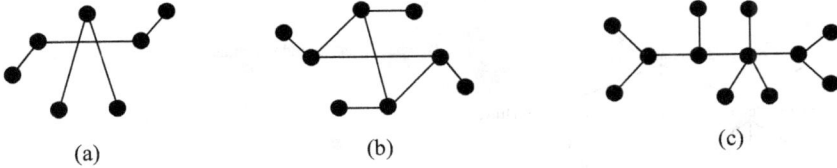

(a)   (b)   (c)

Figure 3.7

Example 3.2.1. *(1) Every star $K(1, q)$, where $q \geq 1$, is a tree. It is noted that $d(u, v) \leq 2$ for any two vertices u, v in $K(1, q)$.*

*(2) A graph of order n is called a **path** if its vertices can be named as $v_1, v_2, \cdots, v_n$ in such a way that v_i is adjacent to v_{i+1} for each $i = 1, 2, \cdots, n - 1$ and no other adjacency exists. Such a path of order n is denoted by P_n. Note that $P_1 \cong K_1$, $P_2 \cong K_2$, and for $n \geq 3$, P_n can be obtained from C_n by deleting an edge in C_n. Clearly, P_n is a tree with exactly two end-vertices, namely v_1 and v_n, and $d(v_1, v_n) = n - 1$. The graph P_7 is shown in Figure 3.8.*

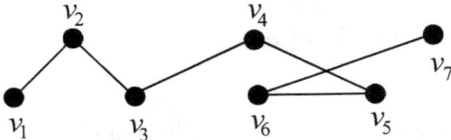

Figure 3.8

Recall that in Section 2.2, a 'path' is introduced as a special walk in a multigraph, but now a **path** *is regarded as a graph. What a 'path' really means should be clear from the context when it is mentioned.*

(3) Some other examples of trees are shown in Figure 3.9.

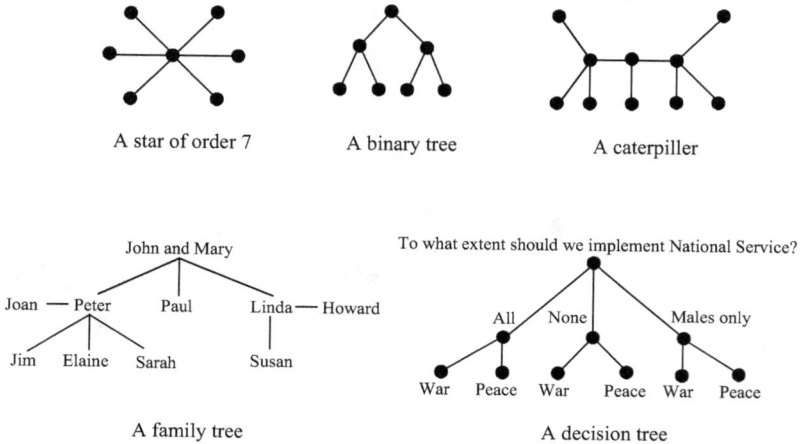

A star of order 7   A binary tree   A caterpiller

A family tree   A decision tree

Figure 3.9

3.2.1 *Special properties for trees*

A tree is a very special graph, and so it must have certain properties that other graphs do not have. In what follows, we shall find out some of them.

We notice that the trees in Figures 3.7, 3.8 and 3.9 all have at least two end-vertices. Actually this property holds for every tree of order at least two.

Theorem 3.3. *Every tree T of order at least two has at least two end-vertices.*

Proof. This property follows from the following claim.

Claim If P is a longest path in T, then both ends of P are of degree 1.

Let $P : v_1 v_2 \cdots v_k$ be a longest path in T. Since the order of T is at least 2, we have $k \geq 2$.

Suppose that $d_T(v_1) \geq 2$. Then v_1 has a neighbour, say w, which is different from v_2. Since P is a longest path in T, w must be in path P, i.e., $w \in \{v_3, \cdots, v_k\}$, implying that T has a cycle. It is a contradiction to the definition of trees. Therefore, $d_T(v_1) = 1$.

Similarly, it can be shown that $d_T(v_k) = 1$. Thus, the claim holds. $\square$

Question 3.2.2. *Consider the two connected graphs in Figure 3.10. Observe that graph (a) is a tree while graph (b) is not.*

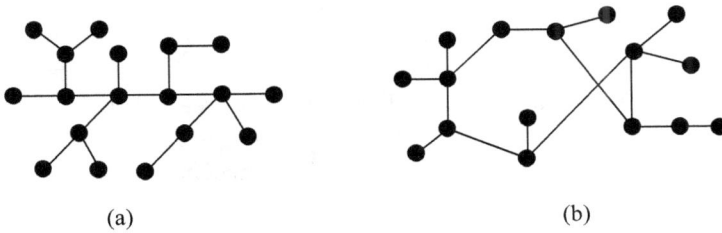

(a)   (b)

Figure 3.10

(i) Is it true that every two vertices in (a) are joined by exactly one path?

(ii) Is it true that every two vertices in (b) are joined by exactly one path?

A tree is connected, and thus it is not surprising that every two of its vertices are joined by a path. An important feature of trees that other connected graphs do not have is the *uniqueness* of such a path joining any two vertices.

Theorem 3.4. *Let G be a connected graph. Then G is a tree if and only if every two vertices in G are joined by a unique path.*

Proof. [Necessity] Suppose that G is a tree and assume on the contrary that there exist two distinct vertices x, y in G which are joined by two different $x - y$ paths, say P and Q. Denote by u the first vertex (as we traverse from x to y) on both P and Q such that its successors on P and Q are distinct. Denote by v the next vertex which lies on both P and Q (note that both u and v exist; at one extreme $u = x$ and $v = y$) (see Figure 3.11). It is now clear that the union of the two $u - v$ paths on P and Q forms a cycle in G, a contradiction.

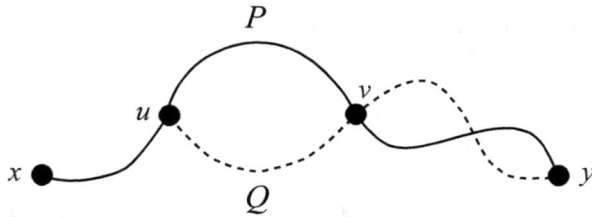

Figure 3.11

[Sufficiency] Suppose that every two vertices in G are joined by a unique path and assume on the contrary that G is not a tree. Then, by definition, G contains a cycle C. Take any two distinct vertices x and y on C. Clearly, x and y are joined by two paths along C in G, a contradiction.   □

Question 3.2.3.   *(i) For each of the trees in Figures 3.6, 3.7(c), 3.8, 3.9, and 3.10(a), find its order and size. Do you notice any relationship between the order and the size?*

(ii) For each connected graph shown in Figures 3.7(b) and 3.10(b), find its order and size. Do they have the same relationship as the one you have found in (i)?

Theorem 3.5. *Let G be a connected graph of order n and size m. Then G is a tree if and only if $m = n - 1$.*

Proof. [Necessity] We prove it by induction on n. The result is trivial if $n = 1$. Assume that the result is true for all trees of order less than n, where $n \geq 2$, and let G be a tree of order n. Choose an edge, say xy, in G. By Theorem 3.4, xy is the unique $x - y$ path in G. Thus, $G - xy$ is a disconnected graph having two components, say G_1 and G_2, both of which are also trees. Let n_i and m_i be, respectively, the order and size of G_i, $i = 1, 2$. By the induction hypothesis, we have $m_i = n_i - 1$ for each $i = 1, 2$. Thus,

$$m = m_1 + m_2 + 1 = (n_1 - 1) + (n_2 - 1) + 1 = n_1 + n_2 - 1 = n - 1,$$

as was to be shown.

[Sufficiency] Assume that G is connected and $m = n - 1$.

We shall now prove that G is a tree by induction on n. The result is trivial if $n = 1$. Assume that $n \geq 2$. By Theorem 3.3, G contains an end-vertex, say w. Clearly, $G - w$ is connected, and $e(G - w) = v(G - w) - 1$. By the induction hypothesis, $G - w$ is a tree. It follows that G is a tree, as desired.   □

By Theorem 3.3, we know that every tree of order at least 2 has at least two end-vertices. We have also pointed out that every path P_n, $n \geq 2$, has exactly two end-vertices. Indeed, we find more end-vertices in other trees shown above.

Question 3.2.4.   *Look at the tree T of Figure 3.12.*

T:

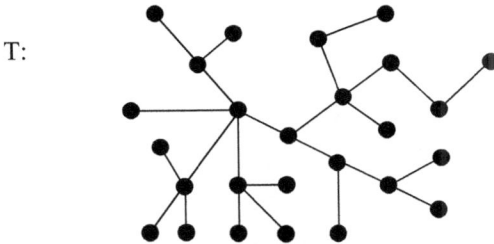

Figure 3.12

For convenience, let us denote by n_i the number of vertices of degree 'i' in T, where $i = 1, 2, \cdots$. Thus n_1 counts the number of end-vertices in T.

(i) Find $\Delta(T)$.

(ii) Count n_i, $i = 1, 2, 3, 4, 5$.

(iii) Evaluate the sum: $2 + n_3 + 2n_4 + 3n_5$.

(iv) Is your sum obtained in (iii) equal to n_1?

Indeed, we have in general the following interesting result which expresses the number of end-vertices of a tree in terms of the numbers of vertices of degree 3, 4, and so on, up to $\Delta(T)$. The reader is strongly encouraged to prove it (see Problem 14 of Exercise 3.2).

Theorem 3.6. *Let T be a tree having n_i vertices of degree i, where $i =$*

$1, 2, \cdots, k$ *with* $k = \Delta(T)$. *Then* $n_1 = 2 + n_3 + 2n_4 + 3n_5 + \cdots + (k-2)n_k$.

Remarks.

((i)) It follows from Theorem 3.6 that every tree of order $n \geq 2$ contains at least two end-vertices. (Which trees have exactly two end-vertices? See Problem 3 of Exercise 3.2.)

((ii)) The number n_2 is not involved in the above expression. That is, the number of end-vertices in a tree is independent of the number of the vertices of degree two.

Exercise 3.2

(1) Draw all non-isomorphic trees of order n, where $2 \le n \le 6$.

(2) Let G be a graph of order n and size $n - 1$, where $n \ge 4$. Must G be a tree?

(3) Let T be a tree of order $n \ge 2$. Show that T has exactly two end-vertices if and only if T is a path, i.e., $T \cong P_n$.

(4) Let T be a tree of order $n \ge 3$.

   (i) What is T if $d(x, y) \le 2$ for any two vertices x, y in T?
   (ii) What is T if $d(u, v) = n - 1$ for some vertices u, v in T?

(5) Find all trees T of order $n \ge 2$ such that $\overline{T}$ is a tree. Is there any tree of order $n \ge 2$ which is self-complementary?

(6) A connected graph is said to be **unicyclic** if it contains one and only one cycle as a subgraph.

   (i) Is every cycle unicyclic?
   (ii) Construct two unicyclic graphs of order 8 which are not C_8.
   (iii) How many edges are there in each of your graphs in (ii)?

(7) Let G be a unicyclic graph.

   (i) What is the relation between $e(G)$ and $v(G)$? Justify your answer.
   (ii) Show that there exist at least three edges e in G such that $G - e$ is a tree.

(8) (+) Let G be a unicyclic graph and let n_1 denote the number of end-vertices in G. Find an expression for n_1 similar to that in Theorem 3.6.

(9) Let G be a connected graph. Show that G is a tree if and only if every edge in G is a bridge. (See Problem 16 of Exercise 2.3.)

(10) (+) Let T be a tree of order k. Show that if G is a graph with $\delta(G) \ge k - 1$, then T is isomorphic to some subgraph of G.

(11) Let T be a tree of order 15 such that $1 \le d(v) \le 4$ for each vertex v in T. Suppose that T contains exactly 9 end-vertices and exactly 3 vertices of degree 4. How many vertices of degree 3 does T have? Justify your answer. Construct one such tree T.

(12) The degrees of the vertices of a tree T of order 18 are $1, 2$ and 5. If T has exactly 4 vertices of degree 2, how many end-vertices does G have?

(13) Let T be a tree and let n_i be the number of vertices of degree i in T. Which of the following statements is/are true?

(i) If T is not a path, then $n_1 \geq n_2$.

(ii) If $n_2 = 0$, then T has more end-vertices than other vertices.

(14) (+) Let T be a tree having n_i vertices of degree i, where $i = 1, 2, \cdots, k$, with $k = \Delta(T)$. Show that $n_1 = 2 + n_3 + 2n_4 + 3n_5 + \cdots + (k-2)n_k$.

(15) (+) Let T be a tree of order n. Show that the vertices in T can always be named as $x_1, x_2, \cdots, x_n$ so that every x_i has one and only one neighbour in $\{x_1, x_2, \cdots, x_{i-1}\}$, for $i = 2, 3, \cdots, n$.

(16) Let G be a graph of order n with degree sequence $(d_1, d_2, \cdots, d_n)$. Show that if G is a tree, then

$$\sum_{i=1}^{n} d_i = 2(n-1).$$

Is the converse true?

(17) (+) Show that every sequence $(d_1, d_2, \cdots, d_n)$ of positive integers with

$$\sum_{i=1}^{n} d_i = 2(n-1)$$

is a degree sequence of a tree.

(18) A *forest* is a graph which contains no cycle as a subgraph.

(i) Is it true that every tree is a forest?

(ii) Is it true that every forest is a tree?

(iii) Is it true that every component of a forest is a tree?

(iv) (+) Let F be a forest. Find a relation linking $v(F), e(F)$ and $c(F)$, and prove your result.

(19) (+) Let G be a graph of order n and size $n-1$. Prove that G is connected if and only if G contains no cycles.

3.3   Spanning trees of a graph

Consider the graph G of Figure 3.13(a). It is connected. Indeed, it remains connected even if some edges are removed. For instance, as shown in Figure 3.13(b), $T = w - \{uv, xy\}$ is still connected. However, we now cannot afford to miss any edge from T to maintain the connectedness. Observe that T is both (1) a spanning subgraph of G and (2) a tree. It is called a spanning tree of G.

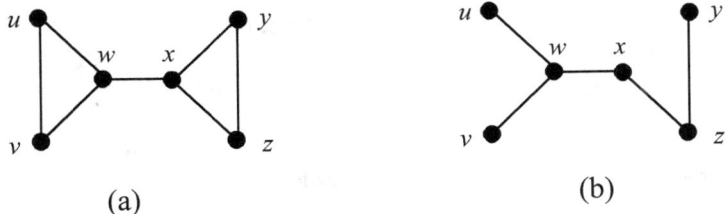

Figure 3.13

In general:

A graph H is called a **spanning tree** of a graph G if H is both (1) a spanning subgraph of G and (2) a tree.

Two more spanning trees of G in Figure 13.3(a) other than T are shown in Figure 3.14.

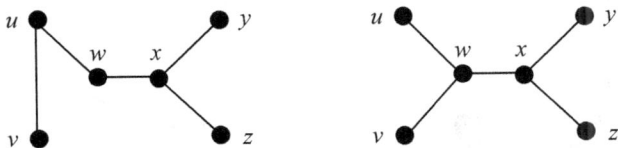

Figure 3.14

Question 3.3.1. *Find all spanning trees of the graph G in Figure 3.13(a). How many are there?*

The existence of a spanning tree of a graph G is directly linked to the connectedness of G as shown below.

Theorem 3.7. *Let G be a graph. Then G is connected if and only if G contains a spanning tree.*

Proof. [Sufficiency] Suppose that G contains a spanning tree, say T. We shall show that G is connected by showing that every two vertices in G are joined by a path in G. Thus, let x and y be any two vertices in G. Then, as T is spanning, x and y are in T. Since T is connected, there is a $x - y$ path in T. As T is a subgraph of G, this $x - y$ path is also in G, as required.

[Necessity] Suppose now that G is connected. If G contains no cycles, then G is itself a spanning tree of G. Otherwise, let C be a cycle of G and e be an edge in C. Then $G - e$ is still spanning and connected (why?). If $G - e$ contains no cycles, then $G - e$ is a spanning tree of G. Otherwise, we proceed as before by deleting an edge from an existing cycle. We continue this procedure repeatedly until a spanning tree of G is eventually found after a finite number of steps.   □

If G is connected graph, then, by Theorem 3.7, G contains a spanning tree T as a subgraph. Thus, by Theorem 3.5, we have:

$$e(G) \geq e(T) = v(T) - 1 = v(G) - 1.$$

Corollary 3.8. *If G is a connected graph of order n and size m, then*

$$m \geq n - 1.$$

Remark. The notion of spanning trees can indeed be defined for multi-graphs, and Theorem 3.7 and Corollary 3.8 remain valid if G is a multi-graph.

Question 3.3.2. *(i) If G is a graph of order n and size m such that $m \geq n - 1$, must G be connected?*
    (ii) If H is a graph of order 100 and size 98, can H be connected? Why?

We shall now give an example to show an application of Corollary 3.8.

Example 3.3.1. *Let G be a connected bipartite graph with a bipartition (X, Y). Assume now that $d(x) \leq 5$ for each x in X. As G is connected and each vertex in X has degree at most five, it is clear that $|Y|$ cannot be too large as compared to $|X|$. It is thus reasonable to ask: what is the best upper bound for $|Y|$ in terms of $|X|$?*

Let us consider $e(G)$. By Result (1) in Section 3.2,

$$e(G) = \sum_{x \in X} d(x).$$

As $d(x) \leq 5$ for each x in X,

$$e(G) = \sum_{x \in X} d(x) \leq 5|X|.$$

On the other hand, as G is connected, by Corollary 3.8,

$$e(G) \geq v(G) - 1 = |X| + |Y| - 1.$$

Now combining the above two inequalities through $e(G)$, we have:

$$|X| + |Y| - 1 \leq e(G) \leq 5|X|$$

and so

$$|Y| \leq 4|X| + 1.$$

That is, Y can have at most $4|X| + 1$ vertices.

To show that the above inequality is sharp (i.e., the equality can be attained), the reader is invited to construct such a bipartite graph with $|Y| = 4|X| + 1$ and $|X| = 1, 2, \cdots$.

Graphs, in particular trees, can be used to model systems as diverse as oil pipelines and internet search programs. The following describes one application of trees in mechanical engineering.

An application. Figure 3.15 shows a 3-D folded structure and one of its flat layouts.

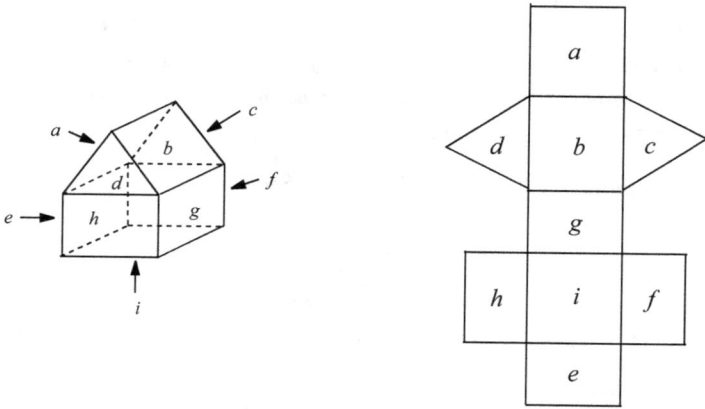

Figure 3.15

Let us introduce a graph to study the relationship between the 3-D folded structure and its flat layout. The 3-D folded structure has nine faces as indicated. Construct a graph G with $V(G) = \{a, b, c, d, e, f, g, h, i\}$, where each vertex represents a face, such that two vertices are adjacent in G if and only if the faces they represent have one edge in common. This graph G, called the **face adjacency graph** (FAG) of the 3-D folded structure, is shown in Figure 3.16.

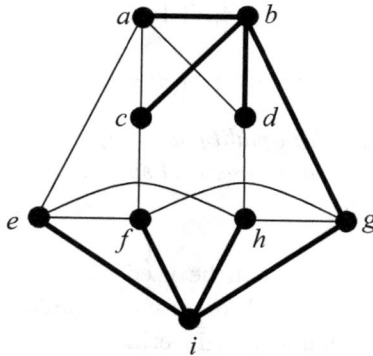

Figure 3.16

If we construct the FAG of the flat layout of Figure 3.15, then we obtain a graph, which is actually a spanning tree of G. In Figure 3.16, this spanning tree of G is shown with the bold edges. In the study of unfolding of 3-D folded structures, one fundamental problem is: given a 3-D folded structure, what are the flat layouts that could be folded into the structure? It is obvious from the above discussion that this problem is actually the problem of finding the spanning trees of the FAG of the given 3-D structure. On the other hand, some 3-D folded structures are 'non-manifold'. Not all the spanning trees of the FAG of such a non-manifold structure give rise to feasible flat layouts of the structure. The challenge here is to find out which spanning trees are feasible and to develop efficient algorithms for the search.

Exercise 3.3

(1) Find all spanning trees of the following graph.

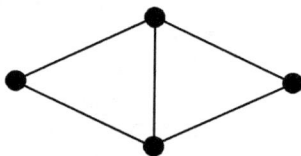

(2) Let H be a graph of order 1000 and size 998. Can H be connected? Why?

(3) Let G be a connected graph and e be a bridge in G. Must e be contained in any spanning tree of G? Why?

(4) Let G be a unicyclic graph which contains C_k as a subgraph, where $k \geq 3$. How many spanning trees does G have?

(5) (+) Prove that every graph of order n and size $n - r$ has at least r components.

(6) Let G be a connected bipartite graph with bipartition (X, Y). Assume that $d(x) \leq 7$ for each x in X. Show that

$$|Y| \leq 6|X| + 1.$$

For each $|X| = 1, 2, \cdots$, construct one such bipartite graph G with $|Y| = 6|X| + 1$.

(7) Let G be a connected bipartite graph with bipartition (X, Y). Assume that G is not a tree and $d(x) \leq 4$ for each x in X. Find the best upper bound for $|Y|$, in terms of $|X|$. Justify your answer.

Chapter 4

Vertex-colourings of Graphs

4.1 The four-colour problem

Figure 4.1 shows a map with a number of regions. One wishes to colour the regions in such a way that adjacent regions (i.e., regions sharing some common boundary) are coloured by different colours. What is the minimum number of colours needed?

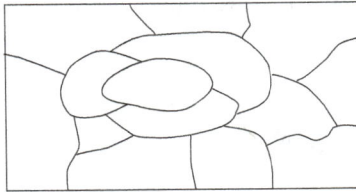

Figure 4.1

Figure 4.2 shows a way of colouring the above map using four colours.

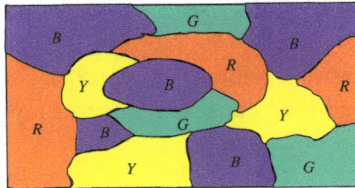

Figure 4.2

Can all maps be coloured with at most four colours? Many people believed that the answer is affirmative, but no one could prove it for a long time. This is known as the **four-colour problem**.

The history of the four-colour problem started in 1852 with an English man named Francis Guthrie (1831 – 1899), who found this fact while colouring regions on a map of England. He told his younger brother Frederick about it, and Frederick, in turn, mentioned it to his teacher Augustus De Morgan (1806 – 1871). De Morgan was then a very well-known professor of mathematics at what is now University College, London, and quite often, he spoke of it to other mathematicians, which certainly helped spread the problem. The problem received further attention after it was formally asked by another prominent English mathematician Arthur Cayley (1821 – 1895) at a meeting of the London Mathematical Society on June 13, 1878.

Amongst many mathematical problems, perhaps the four-colour problem was one of the most famous ones. Many researchers had claimed that they had proved it, but their 'proofs' always contained some fatal flaw somewhere.

It was not until fifty years ago that the problem was eventually settled. In the summer of 1976, Kenneth Appel and Wolfgang Haken, two professors at the University of Illinois, USA, announced that they had solved the problem affirmatively. Their approach was to divide the problem into nearly 2,000 cases and then to write computer programs to analyze the various cases, which required more than 1,200 hours of computer calculations. Twenty years later, a new and simpler proof (but still computer-assisted) was provided by Robertson, Sanders, Seymour and Thomas (see [RSST]). For more information about the history and development of the problem, the reader is referred to [W].

4.2 Vertex-colourings and chromatic number

Studying the four-colour problem by map drawings and real colourings could be quite messy and time consuming if the maps are complicated. Let us instead use 'graph' as a model.

For a given map such as that in Figure 4.1, we obtain a graph by representing each region by a vertex, and joining two vertices by an edge if and only if the two corresponding regions are adjacent. The graph corresponding to the map of Figure 4.1 is shown in Figure 4.3.

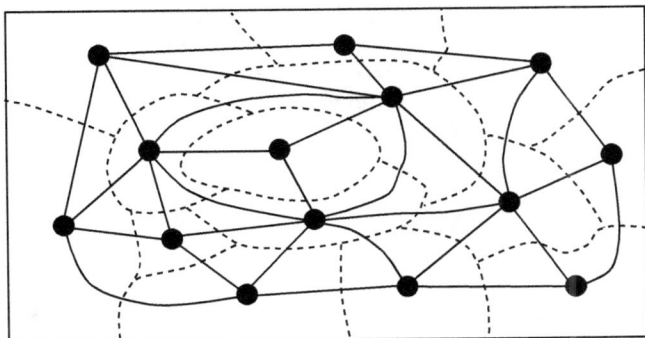

Figure 4.3

Now, 'colouring regions' in a map means 'colouring vertices' in its corresponding graph. Instead of painting the regions by real colours such as blue, green, red, yellow, etc., let us make it more simple by colouring the vertices with colours 1, 2, 3, 4 and so on. Thus the colouring of the vertices of the graph in Figure 4.3 corresponding to the colouring of the regions in Figure 4.2 is now shown in Figure 4.4. Note that adjacent vertices in the graph are coloured by different colours.

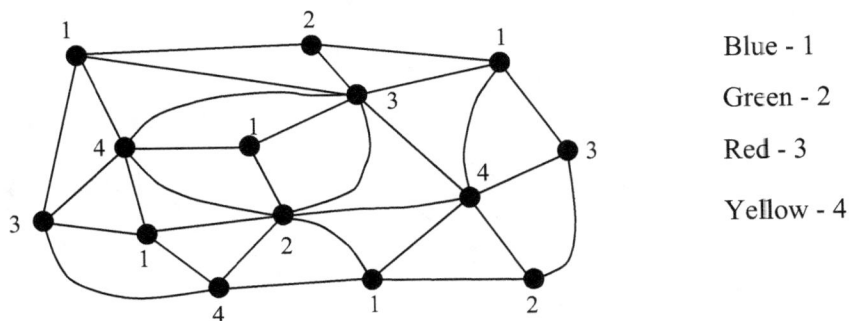

Blue - 1

Green - 2

Red - 3

Yellow - 4

Figure 4.4

The above discussion motivates us to introduce the following notion of 'vertex-colouring' for a general graph.

Let G be a graph and k a positive integer. A k-**colouring** of G is a way of colouring the vertices in G by **at most** k colours in such a way that adjacent vertices are coloured by different colours.

Mathematically, a k-colouring of G can be regarded as a mapping

$$\theta : V(G) \to \{1, 2, \cdots, k\}$$

(not necessarily onto) such that $\theta(u) \neq \theta(v)$ if the vertices u and v are adjacent in G.

Example 4.2.1.   *Consider the graph G of Figure 4.5.*

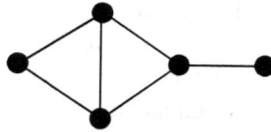

Figure 4.5

A 5-colouring, a 4-colouring and a 3-colouring of G are, respectively, shown in (a), (b) and (c) of Figure 4.6.

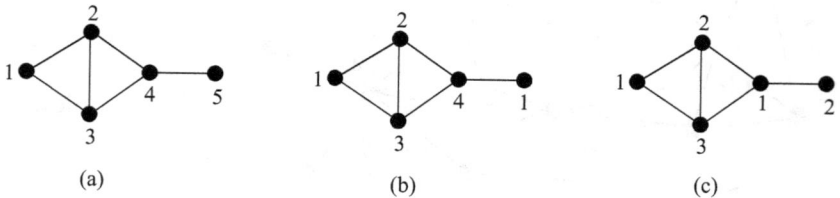

Figure 4.6

Remark. By definition, a p-colouring of G is also a q-colouring of G if $p \leq q$. Thus, it is absolutely correct to say that the three colourings in Figure 4.6 are 5-colourings of G.

Question 4.2.1. *Let G be the graph of Figure 4.7.*

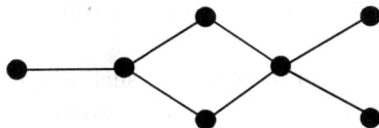

Figure 4.7

(i) Find a 4-colouring, a 3-colouring and a 2-colouring of G.

(ii) Does there exist a 1-colouring of G? Why?

Question 4.2.2. *Let G be the graph of Figure 4.8.*

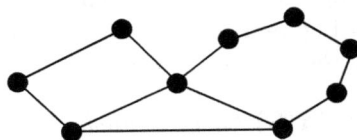

Figure 4.8

(i) Find a 5-colouring, a 4-colouring and a 3-colouring of G.

(ii) Does there exist a 2-colouring of G? Why?

Example 4.2.2. *For a graph G, let k be the smallest integer such that G admits a k-colouring.*

(i) What is the value of k if G is the graph in Figure 4.7?

(ii) What is the value of k if G is the graph in Figure 4.8?

(iii) What is the value of k if G is the graph in Figure 4.5?

Solution.   (i) In Question 4.2.1, you have provided a 2-colouring of G of Figure 4.7; and it is clear that G admits no 1-colourings (since G contains an edge). Thus, $k = 2$.

(ii) In Question 4.2.2, you have provided a 3-colouring of G of Figure 4.8; and it has been discussed that G admits no 2-colourings (since G contains a triangle). Thus, $k = 3$.

(iii) In Figure 4.6(c), we provide a 3-colouring of G of Figure 4.5; and as G contains a triangle, G admits no 2-colourings. Thus, $k = 3$. □

The above discussion on determining such a smallest 'k' motivates us to introduce the following 'key' notion on vertex-colourings.

Let G be a graph. The **chromatic number** of G, denoted by $\chi(G)$, is the minimum value of k such that G admits a k-colouring. That is, $\chi(G)$ is the smallest number of colours needed to colour the vertices of G in such a way that adjacent vertices are coloured by different colours.

Thus, if G is the graph of Figure 4.5, then $\chi(G) = 3$;
if G is the graph of Figure 4.7, then $\chi(G) = 2$; and
if G is the graph of Figure 4.8, then $\chi(G) = 3$.

Question 4.2.3.   *Find $\chi(G)$ for each graph G in Figure 4.9.*

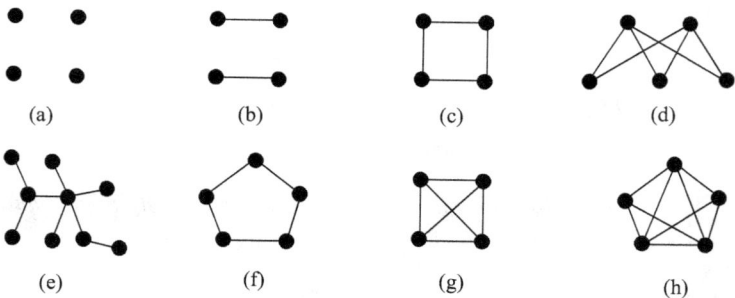

Figure 4.9

Question 4.2.4.   *Let G be a graph of order n. What is the least possible value for $\chi(G)$? What is the largest possible value for $\chi(G)$?*

Question 4.2.5.   *If H is a subgraph of a graph G, what is the relation between $\chi(H)$ and $\chi(G)$?*

Remark. There is another convenient way to express the fact that there exists a k-colouring of a graph. A graph G is said to be **k-colourable** if G admits a k-colouring. Thus, the chromatic number $\chi(G)$ is the smallest k such that G is k-colourable. Using this terminology, then the four-colour problem (theorem) states that if G is the graph obtained from a map of regions as shown in Figure 4.3, then G is 4-colourable, that is, $\chi(G) \leq 4$.

Exercise 4.2

(1) Consider the following map:

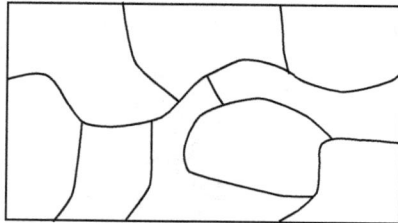

  (i) Colour the regions with no more than four colours in such a way
      that each region is coloured by one colour, and adjacent regions are
      coloured by different colours.
 (ii) Construct a graph G modeling the above situation as shown in Fig-
      ure 4.3.
(iii) Does G contain a K_4 as a subgraph?
 (iv) Does G contain a K_5 as a subgraph?
  (v) Is G 3-colourable? Why?
 (vi) Is G 4-colourable?
(vii) What is the value of $\chi(G)$?

(2) Let p and q be integers such that $1 \le p \le q$. Explain by definition why
    a p-colouring of a graph is also a q-colouring of the graph.

(3) Prove that if H is a subgraph of a graph G, then $\chi(H) \le \chi(G)$.

(4) Construct two graphs H and G such that H is a proper subgraph of G
    but $\chi(H) = \chi(G)$.

(5) Construct two connected graphs H and G such that H is a spanning
    subgraph of G and $\chi(H) = \chi(G) - 1$.

(6) For each of the following graphs, find its chromatic number.

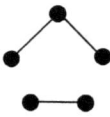
(a)
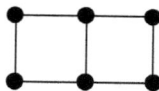
(b)
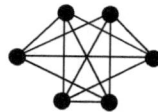
(c)

(d)

4.3 Enumeration of chromatic number

Throughout this section, let G be a graph of order n. From Question 4.2.3 you have probably come to the conclusion that $1 \leq \chi(G) \leq n$ (see also Question 4.2.4).

Which graphs G have their $\chi(G)$ equal to 1? Which graphs G have their $\chi(G)$ equal to its order $v(G)$?

Example 4.3.1. *For $n \geq 1$, $\chi(N_n) = 1$ and $\chi(K_n) = n$.*

Whenever there is an edge in G, its two ends are adjacent and must be coloured by different colours. Thus, if $G \not\cong N_n$ for $n \geq 2$ (that is, $e(G) \geq 1$), then $\chi(G) \geq 2$. We have:

Result (1). Let G be a graph of order n. Then $\chi(G) = 1$ if and only if $G \cong N_n$.

Whenever there are two vertices in G which are not adjacent, we may colour these two vertices by the same colour, and so we use at most $n - 1$ colours to colour G. Thus, if $G \not\cong K_n$ for $n \geq 2$, then $\chi(G) \leq n - 1$. We have:

Result (2). Let G be a graph of order $n \geq 2$. Then $\chi(G) = n$ if and only if $G \cong K_n$.

Now, which graphs G have their $\chi(G)$ equal to 2?

Example 4.3.2. *Let G be a bipartite graph with bipartition (X, Y) which contains at least one edge. Since no edge joins two vertices in X, we can colour the vertices in X by one colour. Likewise, we can colour the vertices in Y by one colour. As there is an edge joining a vertex in X to a vertex in Y, the colour for X must be different from that for Y. Thus, $\chi(G) = 2$.*

Remark. The null graphs N_n and all trees with at least two vertices are, by definition, bipartite graphs.

Question 4.3.1.   *If $\chi(G) = 3$, is it true that G contains an odd cycle as a subgraph?*

The result in Example 4.3.2 says that if G is a bipartite graph with at least one edge, then $\chi(G) = 2$. Is the converse true? That is, if G contains at least one edge and $\chi(G) = 2$, must G be bipartite?

Since bipartite graphs and odd cycles are closely related by Theorem 3.1, to answer this question, we first evaluate $\chi(C_n)$.

Example 4.3.3.   *For $n \geq 3$, $\chi(C_n) = \begin{cases} 2, & \text{if } n \text{ is even;} \\ 3, & \text{if } n \text{ is odd.} \end{cases}$*

In the discussion of Question 4.2.5, we learn that if H is a subgraph of a graph G, then $\chi(H) \leq \chi(G)$. Now, if G is not bipartite, then by Theorem 3.1, G contains an odd cycle C_{2k+1} as a subgraph, and so $\chi(G) \geq \chi(C_{2k+1}) = 3$, by Example 4.3.3. It thus follows that if $\chi(G) = 2$, then G must be bipartite; which says that the converse of the result in Example 4.3.2 is true. We now re-state this important observation as follows:

Result (3). Let G be a graph with at least one edge. Then $\chi(G) = 2$ if and only if G is bipartite.

Question 4.3.2.   *Construct all graphs G of order n, $2 \leq n \leq 5$, such that $\chi(G) = n - 1$. (See Problem 7 of Exercise 4.3.)*

We have determined all graphs G with $\chi(G)$ equal to 1 or 2. The problem of characterizing the graphs G with $\chi(G) = 3$ remains unsettled.

However, from our discussion before Result (3), we do have:

Result (4). Let G be a graph which contains an odd cycle as a subgraph. Then $\chi(G) \geq 3$.

Likewise, we have:

Result (5). Let G be a graph and let p be any positive integer such that G contains a K_p as a subgraph. Then $\chi(G) \geq p$.

Remark. To get a sharper lower bound in Result (5), of course, we look for the largest such p.

Example 4.3.4. *Determine the chromatic number of the graph G in Figure 4.10.*

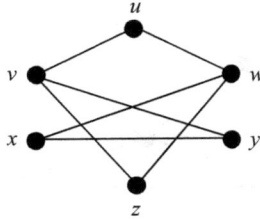

Figure 4.10

Since G contains a C_5 : uvyxwu, by Result (4), $\chi(G) \geq 3$. As a 3-colouring of G is given in Figure 4.11, we conclude that $\chi(G) = 3$.

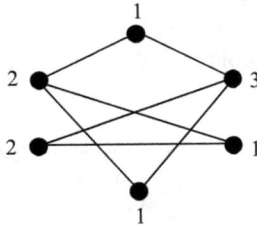

Figure 4.11

Example 4.3.5. *Consider the graph G of Figure 4.4. There are many K_3's in G. Is there any K_4 in G? Yes, there is one (try to find it!). Is there any K_5 in G? Definitely 'no'! Now, by Result (5), $\chi(G) \geq 4$. As a 4-colouring of G is shown in Figure 4.4, we thus conclude that $\chi(G) = 4$.*

Example 4.3.6.   *Determine the chromatic number of the graph G in Figure 4.12.*

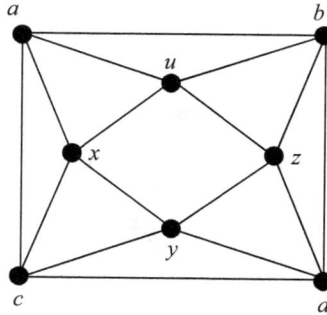

Figure 4.12

Since G contains a triangle, $\chi(G) \geq 3$. Now, if we could provide a 3-colouring for G, then we would conclude that $\chi(G) = 3$. Let us try whether it is possible.

Thus, assume that G admits a 3-colouring using colours: $1, 2$ and 3. Choose a triangle, say $axua$, to begin with. These three vertices must be coloured by three different colours. We may assume the following:

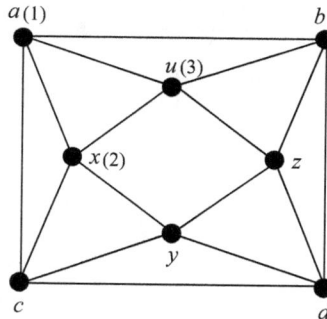

Consider the triangle $axca$, say. Then the vertex c must be coloured by 3. Proceed in this manner as shown in the following sequence:

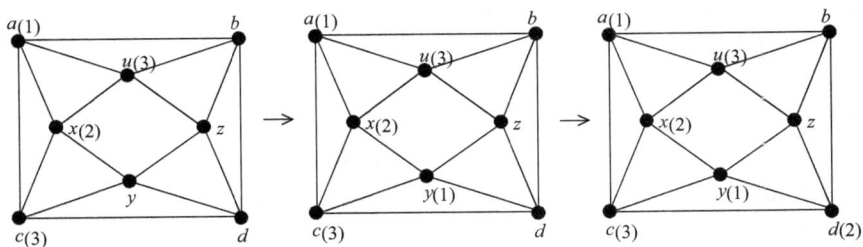

We now look at the vertex z. Observe that three of its neighbours are coloured by colours: $1, 2$ and 3. We are now forced to use a 'new' colour to colour z by definition of colourings.

The above argument shows that three colours are not enough to colour the graph. Thus, $\chi(G) \geq 4$.

As a 4-colouring of G is shown in Figure 4.13, we eventually conclude that $\chi(G) = 4$.

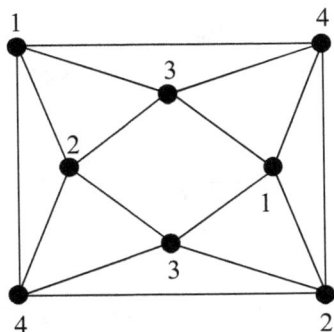

Figure 4.13

Exercise 4.3

(1) Prove that $\chi(C_n) = 2$ for any even $n \geq 4$.

(2) Prove that $\chi(C_n) = 3$ for any odd $n \geq 3$.

(3) Find a 3-colouring of the Petersen graph. What is its chromatic number?

(4) Let G be the graph given below. Explain why $\chi(G) \geq 3$. Then provide a 3-colouring for G, thereby proving that $\chi(G) = 3$.

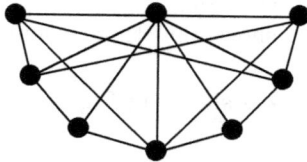

(5) Let G be the graph given below. Explain why $\chi(G) \geq 4$. Then provide a 4-colouring for G, thereby proving that $\chi(G) = 4$.

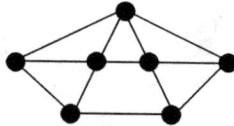

(6) (+) Let G be the graph given below. Explain why $\chi(G) \geq 4$. Then provide a 4-colouring for G, thereby proving that $\chi(G) = 4$.

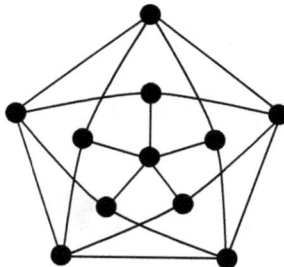

(7) (+)   Determine all graphs G of order $n \geq 2$ with $\chi(G) = n - 1$.

(8) Let G be a graph. Determine whether each of the following statements is true.

   (i) If G admits a 3-colouring, then G is 3-colourable.

  (ii) If G is 3-colourable, then G is 5-colourable.

 (iii) If G is 3-colourable, then $\chi(G) \geq 3$.

 (iv) If G is 3-colourable, then $\chi(G) \leq 3$.

  (v) If G is 3-colourable, then G contains an odd cycle.

 (vi) If G contains an odd cycle, then G is 3-colourable.

(vii) If G admits no 3-colourings, then $\chi(G) \geq 3$.

(viii) If G admits no 3-colourings, then $\chi(G) = 2$.

 (ix) If G admits no 3-colourings, then $\chi(G) \leq 2$.

  (x) If $\chi(G) = 3$, then G contains a triangle.

 (xi) If $\chi(G) = 3$, then G contains an odd cycle.

(xii) If G is a tree with at least two vertices, then $\chi(G) = 2$.

(xiii) If $\chi(G) \geq r$, then G contains a K_r as a subgraph.

(9) Let G be a disconnected graph with two components G_1 and G_2. Show that

$$\chi(G) = \max\{\chi(G_1), \chi(G_2)\}.$$

(10) Let G_1 and G_2 be two connected graphs and let G be the graph obtained from G_1 and G_2 by identifying a vertex in G_1 with a vertex in G_2 as shown below:

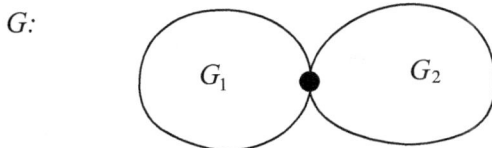

Show that

$$\chi(G) = \max\{\chi(G_1), \chi(G_2)\}.$$

(11) Let G be a graph. Show that

  (i) $\chi(G) - 1 \leq \chi(G - v) \leq \chi(G)$ for each vertex v in G.

 (ii) $\chi(G) - 1 \leq \chi(G - e) \leq \chi(G)$ for each edge e in G.

(12) For each integer $n \geq 2$, construct a graph G of order n such that $\chi(G - v) = \chi(G) - 1$ for each vertex v in G.

(13) For each integer $n \geq 4$, construct a graph G of order n such that $\chi(G - v) = \chi(G)$ for each vertex v in G.

(14) For each integer $n \geq 2$, construct a graph G of order n such that $\chi(G - e) = \chi(G) - 1$ for each edge e in G.

(15) For each integer $n \geq 3$, construct a graph G of order n such that $\chi(G - e) = \chi(G)$ for each edge e in G.

(16) Let G be the graph shown below:

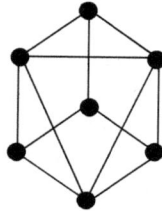

    (i) Find $\chi(G)$.

   (ii) Verify that $\chi(G - e) = \chi(G) - 1$ for each edge e in G.

(17) Construct a graph G such that $\chi(G) = 3$ and G contains no triangles.

(18) Construct a graph G such that $\chi(G) = 4$ and G contains no triangles.

(19) Let H be the graph given below. What is the value of $\chi(H)$?

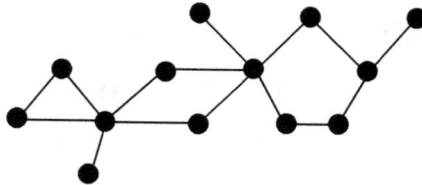

(20) (+) Let G be a graph which contains only one odd cycle as a subgraph. Find the value of $\chi(G)$. Justify your answer.

(21) (+) Let G be a graph which is not bipartite. Assume that there is a vertex in G which is contained in every odd cycle in G. Show that $\chi(G) = 3$.

(22) (+) Let G be a graph. It is known that if $\chi(G) = 3$, then G contains an odd cycle. Assume now that $\chi(G) = 6$. Does G contain two odd cycles which have no vertex in common? Why?

(23) Let G be a graph of order 8 with $\chi(G) = 2$. Show that $e(G) \leq 16$. Construct one such G with $e(G) = 16$.

(24) Let G be a graph of order 7 with $\chi(G) = 3$. Show that $e(G) \leq 16$. Construct one such G with $e(G) = 16$.

(25) Let G be a graph of order 6 with $\chi(G) = 4$. Show that $e(G) \leq 13$. Construct one such G with $e(G) = 13$.

(26) Determine the chromatic number of each of the following graphs:

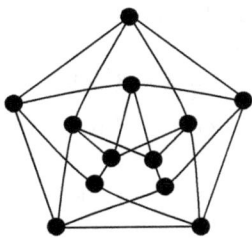   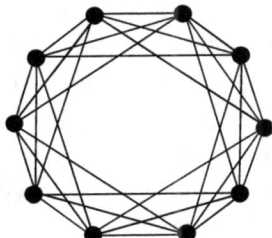

(a)   (b)

(27) Let G and H be two graphs. The **join** of G and H, denoted by $G + H$, is the graph whose vertex set is the union of $V(G)$ and $V(H)$, and whose edge set consists of the edges in G and H together with new edges which join every vertex in G to every vertex in H. Thus, if A and B are the graphs shown in (a) and (b) below, then $A + B$ is the graph in (c).

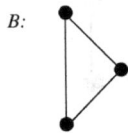   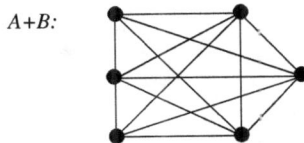

A: B: A+B:

(a)   (b)   (c)

(i) Find $\chi(A), \chi(B)$ and $\chi(A + B)$.
(ii) In general, what is the relation among $\chi(G), \chi(H)$ and $\chi(G + H)$? Prove your result.

(28) The wheel of order n, denoted by W_n, is defined as (see Problem 27 above)

$$W_n = C_{n-1} + K_1.$$

(i) Draw W_6 and W_7.
(ii) Find a 3-colouring for W_7.
(iii) Find a 4-colouring for W_6.
(iv) Show that $\chi(W_n) = 3$ for odd $n \geq 5$.
(v) Show that $\chi(W_n) = 4$ for even $n \geq 4$.

(29) (+) Let G be a graph of order $n \geq 5$ which contains a P_5 as an induced subgraph. Show that $\chi(G) \leq n - 3$. For each $n \geq 5$, construct one such G of order n for which the equality $\chi(G) = n - 3$ holds.

(30) Let G be a graph. A set of vertices S in G is said to be **independent** if no two vertices in S are adjacent.
Assume that $\chi(G) = k$ with a k-colouring θ. For each $i = 1, 2, \cdots, k$, let V_i be the set of vertices v in G with $\theta(v) = i$.

(i) Can V_i be empty?
(ii) Is V_i an independent set?

(31) (+) Let G be a graph of order n. The **independence number** of G, denoted by $\alpha(G)$, is defined by

$$\alpha(G) = \max\{|S| \mid S \text{ is an independent set in } G\}.$$

(i) Find $\alpha(H)$, where H is the graph shown below:

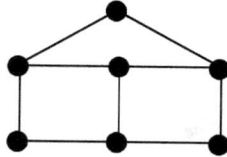

(ii) Show that $\chi(G)\alpha(G) \geq n$.
(iii) Construct a connected graph H such that $v(H) = 12$, $\chi(H) = 4$ and $\alpha(H) = 3$.
(iv) Show that $\chi(G) + \alpha(G) \leq n + 1$.
(v) Construct a connected graph H such that $v(H) = 11$, $\chi(H) = 5$ and $\alpha(H) = 7$.

(32) (+) Let G be a graph which is not bipartite. Assume that G contains an independent set S such that $V(C) \cap S$ is non-empty for every odd cycle C in G. Show that $\chi(G) = 3$.

(33) (+) Let G be a graph satisfying the following conditions:
  (1) $\chi(G) = 5$ and
  (2) $\chi(G - v) = 4$ for each vertex v in G.
  Show that

  (i) G is connected;
  (ii) $\delta(G) \geq 4$;
  (iii) $N(u)$ is not a subset of $N(v)$ for any two vertices u, v in G;
  (iv) $v(G) \neq 6$.

4.4   Greedy colouring algorithm

The problem of evaluating the chromatic number of a graph in general is very difficult; and until now, there is no 'efficient' way to compute its **exact value**. However, there is an 'efficient' heuristic algorithm, called **the greedy colouring algorithm**, which enables us to colour a graph and obtain an approximation to its chromatic number.

Greedy Colouring Algorithm.

Let G be a graph of order n and list its vertices as $v_1, v_2, \cdots, v_n$.

(1) Colour v_1 by colour '1'.

(2) Consider v_2. Find the smallest colour number which is not used to colour the neighbours of v_2. Colour v_2 by this colour number.

(3) Following the ordering: v_i, $i = 3, 4, \cdots, n$, repeat the way of colouring in (2) to colour v_i until all vertices are coloured.

Example 4.4.1.   *Let G be the graph (a) in Figure 4.14. Fix two different orderings of its vertices as shown in Figure 4.14 (b) and (c). We shall apply the greedy colouring algorithm to colour G in two ways following the respective ordering.*

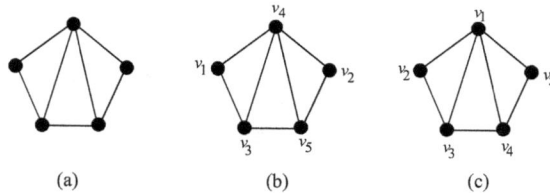

(a)   (b)   (c)

Figure 4.14

Following the ordering shown in (b), the algorithm is applied step by step as shown in the sequence of diagrams:

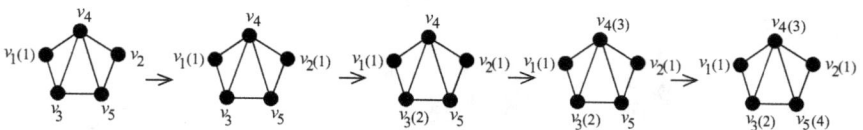

The number of colours produced in this case is '4'.

Now, following the ordering shown in (c), the algorithm is applied step by step as shown in the sequence of diagrams:

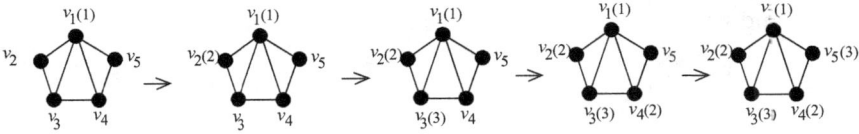

The number of colours produced in this case is '3'.

(1) The above example tells us that the number of colours produced by applying the greedy colouring algorithm depends on the ordering of the vertices.

(2) The number of colours produced by applying the greedy colouring algorithm provides only an upper bound for the chromatic number of the graph. Thus, following the ordering (b), $\chi(G) \leq 4$; while following (c), $\chi(G) \leq 3$.

(3) As G contains a triangle, $\chi(G) \geq 3$. Thus, combining it with the result following (c), we see that $\chi(G) = 3$.

Question 4.4.1. *Let G be the graph shown in Figure 4.15.*

(i) *Arrange its vertices in two different ways and, for each way, apply the greedy colouring algorithm to colour G and to find an upper bound for $\chi(G)$.*

(ii) *Determine $\chi(G)$.*

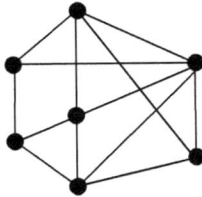

Figure 4.15

Exercise 4.4

(1) Let G be the graph C_6 as shown below with two different ways of arranging its vertices. Apply the greedy colouring algorithm to colour G and find the number of colours produced in each case.

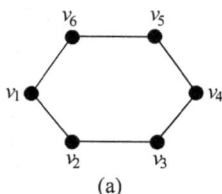

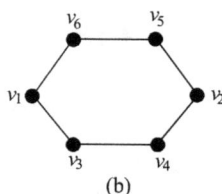

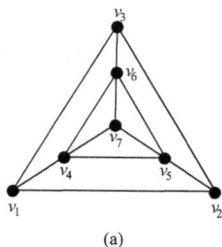

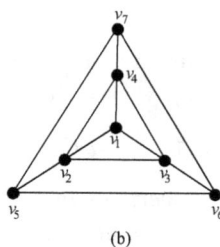

(a)   (b)

(2) Let G be the graph as shown below with two different ways of arranging its vertices. Apply the greedy colouring algorithm to colour G and find the number of colours produced in each case.

(a)   (b)

(3) Let G be the graph given below:

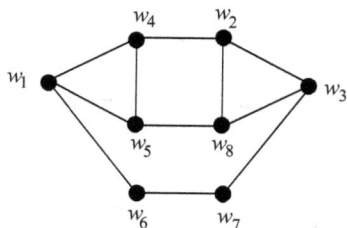

   (i) Find the number of colours produced by applying the greedy colouring algorithm on G according to the ordering of vertices $w_1, w_2, \cdots, w_8$.

 (ii) Find $\chi(G)$.

(4) Let G be the graph given below:

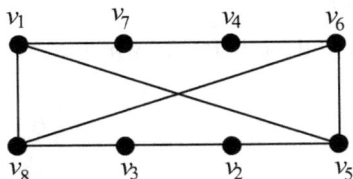

(i) Find the number of colours produced by applying the greedy colouring algorithm on G according to the ordering of vertices $v_1, v_2, \cdots, v_8$.

(ii) Find $\chi(G)$.

(5) Let H be the graph given below:

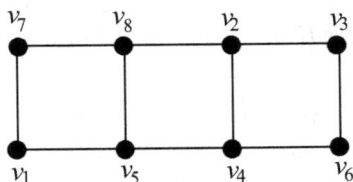

(i) Find the number of colours produced by applying the greedy colouring algorithm on H according to the ordering: $v_1, v_2, \cdots, v_8$.

(ii) Determine $\chi(H)$.

(6) Let H be the graph given below:

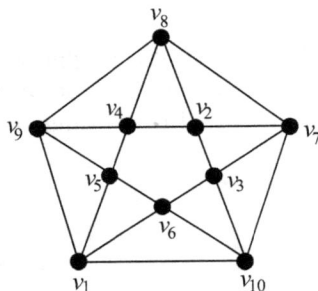

(i) Find the number of colours produced by applying the greedy colouring algorithm on H according to the ordering: $v_1, v_2, \cdots, v_{10}$.

(ii) Determine $\chi(H)$.

(7) Let H be the graph given below in which the vertices are named.

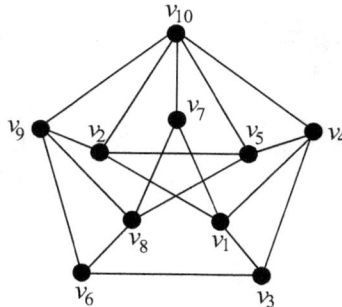

(i) Find the number of colours produced by applying the greedy colouring algorithm on H according to the ordering: $v_1, v_2, \cdots, v_{10}$.

(ii) Determine $\chi(H)$.

(8) (+) Let G be the graph given below.

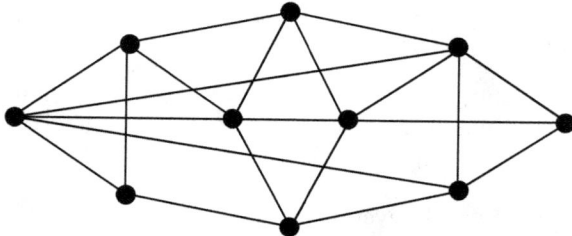

(i) Find $\chi(G)$.

(ii) Arrange the vertices as $v_1, v_2, \cdots, v_{10}$ so that, when the greedy
colouring algorithm is applied to G according to this ordering, the
number of colours produced is the value of $\chi(G)$.

4.5   An upper bound for the chromatic number and Brooks' theorem

As pointed out earlier, it is very difficult to determine the exact value of $\chi(G)$ of a graph G in general. Thus we should be content with some good bounds for $\chi(G)$. While Result (4) and Result (5) in Section 4.3 provide two lower bounds for $\chi(G)$, the greedy colouring algorithm presented in Section 4.4 does provide an upper bound for $\chi(G)$ as shown in Example 4.4.1 and Question 4.4.1.

Question 4.5.1. *Let us re-visit Question 4.4.1 and re-consider the colouring of G of Figure 4.15.*

(i) Find $\Delta(G)$.

(ii) While you applied the greedy colouring algorithm to colour v_i by a colour number in Question 4.4.1(i), was this colour number always in $\{1, 2, \cdots, \Delta(G) + 1\}$? Why?

(iii) Did you need to use more than '$\Delta(G) + 1$' colours to colour G when you applied the greedy colouring algorithm?

Recall the procedure of applying the greedy colouring algorithm to colour the vertices: $v_1, v_2, \cdots, v_n$. In each step of colouring v_i, we first look at its neighbours, then find out the smallest colour number which is not used to colour its neighbours, and finally colour v_i by this 'missing' colour. How many neighbours does v_i have? There are '$d(v_i)$' that many, which is certainly not more than '$\Delta(G)$'. Clearly, the 'worst' case is when the neighbours of v are coloured by '$\Delta(G)$' different colours; but even that, there is still one colour available in $\{1, 2, \cdots, \Delta(G)+1\}$ to colour v_i. Thus, when applying the greedy colouring algorithm to colour G, we never have to use more than $\Delta(G) + 1$ colours. That is, the number of colours produced by applying the algorithm is never more than '$\Delta(G) + 1$'. Now, by the definition of $\chi(G)$, we have:

Theorem 4.1. *For any graph G, $\chi(G) \leq \Delta(G) + 1$.* $\square$

Is the upper bound in Theorem 4.1 sharp? That is, can the equality $\chi(G) = \Delta(G) + 1$ hold for some graph G?
Yes! Take the odd cycle $G = C_{2k+1}$. Clearly, $\chi(G) = 3$ while $\Delta(G) = 2$. Take the complete graph $G = K_n$. Clearly, $\chi(G) = n$ while $\Delta(G) = n - 1$.

And we see that the equality $\chi(G) = \Delta(G) + 1$ holds in these two instances.

Can you find a graph other than an odd cycle or a complete graph for which the equality holds?

Question 4.5.2.   *Let G be the disconnected graph with two components as shown in Figure 4.16.*

(i) Find $\Delta(G)$ and $\chi(G)$.

(ii) Is $\chi(G)$ equal to $\Delta(G) + 1$?

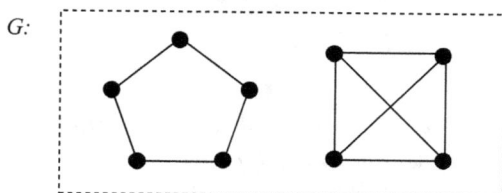

Figure 4.16

Can you find a **connected** graph other than an odd cycle or a complete graph for which the equality holds?

Now, the answer is 'no', and this negative answer was given by R. L. Brooks in 1941.

Theorem 4.2   Let G be a connected graph which is neither an odd cycle nor a complete graph. Then $\chi(G) \leq \Delta(G)$.

Theorem 4.2 is called Brooks' Theorem. In 1975, László Lovász (see [LO]) gave a simple proof of this welll-known result. In the following, we provide a refinement of Lovász's proof.

Proof.   Suppose the result fails, and G is a graph with the minimum order such that Brooks' Theorem fails for G. It is left to readers to verify the following two properties for G:

(a) $\Delta(G) \geq 3$; and

(b) G is 3-connected (i.e., deleting at most two vertices in G cannot disconnect it).

As G is connected and not a complete graph, there must be a path uvw in G such that u and w are not adjacent in G. Since G is 3-connected, $G - \{u, w\}$ is connected and has a spanning tree T'. Then, by adding vertices u and w and edges uv and vw to T', we obtain a spanning tree T of G with the following properties:

(i) uv and vw are edges in T, and

(ii) both u and w are leaves in T (i.e., u and w have degree 1 in T).

For any vertex x in G, let $l(x)$ denote the length of the unique path in T connecting v and x. Thus $l(v) = 0$ and $l(u) = l(w) = 1$. Then, the vertices in G can be labelled as $v_1, v_2, \ldots, v_n$, where $n = |V(G)|$, such that $v_1 = u$, $v_2 = w$, and $l(v_i) \geq l(v_j)$ whenever $3 \leq i < j \leq n$. It follows that v_n is the vertex v. Running the greedy coloring algorithm according to the vertex ordering $v_1, v_2, \ldots, v_n$ assigns color 1 to both v_1 and v_2 (i.e., u and w), ends at v and uses at most Δ colors. For, when vertex v_i is colored, where $3 \leq i \leq n-1$, it has an uncolored neighbour v_j, where $i < j \leq n$, and so its already-colored neighbors cannot use up all the free colors in $\{1, 2, \ldots, \Delta\}$, while at v (i.e., v_n), the two neighbors u (i.e., v_1) and w (i.e., v_2) have the same color (i.e., color 1) so again a free color remains for v itself.

It follows that G is Δ-colorable, contradicting the assumption that the theorem fails for G. Hence, the theorem holds. $\square$

Question 4.5.3.   *In each of the following cases, is $\chi(G) = \Delta(G)$?*

(i) G is a path;

(ii) G is an even cycle;

(iii) G is obtained from K_n, $n \geq 3$, by deleting an edge;

(iv) G is the Petersen graph;

(v) $G = K(p, q)$.

Exercise 4.5

(1) Determine the chromatic number of the following graph. Prove your result.

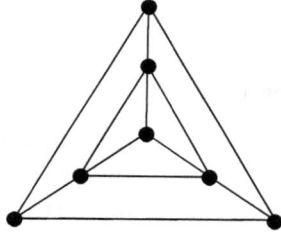

(2) Determine the chromatic number of the following graph. Prove your result.

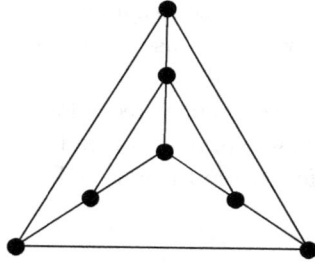

(3) (+) Let G be a connected graph, which is not a complete graph. Show that if G contains K_r as a subgraph, where $r = \Delta(G) \geq 3$, then $\chi(G) = r$.

(4) (+) Consider the following graph H:

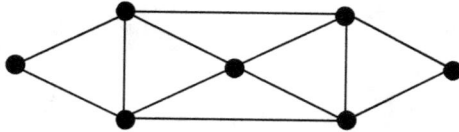

  (i) Find $\Delta(H)$ and $\chi(H)$.

  (ii) Arrange the vertices of H as $v_1, v_2, \cdots, v_7$ so that, when the greedy algorithm is applied to H according to this ordering, the number of colours produced is $\Delta(H) + 1$.

(5) (+) Does there exist a graph G satisfying the following conditions:

   (i) $\chi(G) = 7$ and

   (ii) the degree sequence of G is $(6, 6, 6, 6, 6, 5, 5, 5, 5, 4, 4, 3, 3, 3, 3)$?

(6) (+) Let G be a cubic connected graph. What are the possible values of $\chi(G)$? Classify G according to the value of $\chi(G)$.

(7) (+) Let G be a regular and connected graph of order n. Show that $\chi(G) + \chi(\overline{G}) = n + 1$ if and only if $G \cong K_n$ or $G \cong C_5$.

4.6   Applications

There are a number of applications of vertex colouring. These include the allocation of variables to hardware registers during program execution by a compiler, and the scheduling of traffic lights. In this section, we shall explain in more detail two other applications, namely, scheduling an exam timetable and the storage of combustible chemicals.

Exam Timetable

There are ten students A to J in a postgraduate mathematics programme. At the end of the semester, the students are to take exams for the subjects they are enrolled in. A, B and I took Probability; A, D, I and J took Graph Theory; C, D, F and H took Henstock Integration; E and F took Lie Groups; B, E and G took Non-linear Dynamical Systems; D and J took Topology.

Of course, the exams must be scheduled in such a way that no student has to take two subjects at the same time. In addition, for efficiency, the university would like to minimise the number of timeslots. Can we schedule an exam timetable with a minimum number of timeslots?

To do this, we will draw a graph to model the situation so that we may assign timeslots to the subjects directly on the graph in such a way that no two subjects with a common student are assigned the same timeslot. Here are the questions we will consider:

(1) What will the vertices represent?

(2) Under what condition(s) will two vertices be joined by an edge?

(3) How can we use vertex colouring to assign timeslots to the subjects?

Here are the answers:

(a) Each vertex represents a subject. Thus there will be 6 vertices: Probability (P), Graph Theory (GT), Henstock Integration (H), Lie Groups (L), Non-linear Dynamical Systems (N), Topology (T).

(b) Two vertices (subjects) are joined by an edge if and only if there is a student taking both the subjects.

(c) A vertex colouring of the resulting graph will ensure that no two adjacent vertices will have the same colour. Thus, the colour represents a timeslot and a viable colouring represents a viable exam timetable. The chromatic number of the graph will be the minimum number of timeslots required for a viable timetable.

Let us solve the example above by drawing the graph G to model the situation.

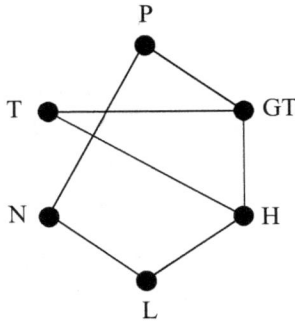

Figure 4.17

As G contains a triangle, $\chi(G) \geq 3$. The 3-colouring of G below shows that indeed $\chi(G) = 3$. Thus, 3 timeslots suffice for a viable exam timetable and we may allocate colour 1 as, say, Monday morning, colour 2 as Tuesday morning and colour 3 as Wednesday morning.

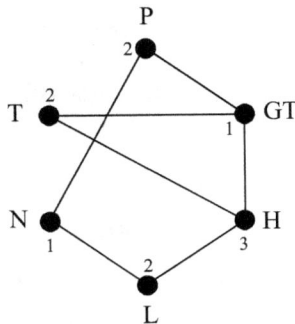

Figure 4.18

Chemical Storage

A common scenario in a factory is the storage of different materials, some of which may be incompatible with others. For example, one would not store gas canisters and matches together.

The case we want to consider here is the storage of chemicals in a factory. Some chemicals when in contact with some other chemicals may result in a combustible or poisonous mixture. These incompatible chemicals ought then to be stored in different rooms. In addition, to optimize the use of space, the factory would like to minimise the number of storerooms needed. Can we allocate the chemicals to a minimum number of storerooms in such a way that chemicals which are combustible together are placed in different rooms?

Let us consider the following concrete example. There are six chemicals U to Z which are used in a factory to manufacture air conditioners. The following table shows which chemicals are incompatible, i.e., they are combustible or poisonous together.

Chemical	Incompatible with:
U	V, Y
V	U, W, Z
W	V, X, Z
X	W, Y
Y	X, U
Z	V, W

Find a safe allocation of the chemicals to a minimum number of rooms.

To do this, we will draw a graph to model the situation so that we may assign rooms to the subjects directly on the graph in such a way that no two incompatible chemicals are assigned the same room. Here are the questions we will consider:

(1) What will the vertices represent?

(2) Under what condition(s) will two vertices be joined by an edge?

(3) How can we use vertex colouring to assign rooms to the chemicals?

Here are the answers:

(a) Each vertex represents a chemical. Thus there will be 6 vertices U to Z.

(b) Two vertices (chemicals) are joined by an edge if and only if they are incompatible.

(c) A vertex colouring of the resulting graph will ensure that no two adjacent vertices will have the same colour. Thus, the colour represents a room and a viable colouring represents a safe allocation of rooms. The chromatic number of the graph will be the minimum number of rooms required for a safe allocation.

Let us solve the example above by drawing a graph H to model the situation.

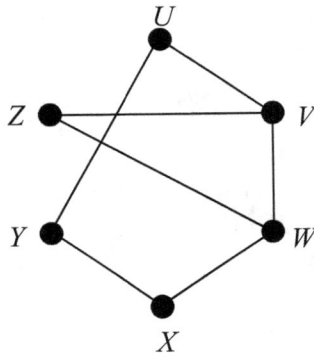

Figure 4.19

As it turns out, the graph H is isomorphic to the graph G in the previous example. Thus, using the argument of the previous example, $\chi(H) = 3$. Hence, 3 rooms suffice for a safe allocation.

Exercise 4.6

(1) A chemist wishes to ship chemicals A, B, C, D, W, X, Y, Z using as few containers as possible. Certain chemicals cannot be shipped in the same container since they will react with each other. In particular, any two of the chemicals in each of the following 6 groups

$$\{A, B, C\}, \{A, B, D\}, \{A, B, X\}, \{C, W, Y\}, \{C, Y, Z\} \text{ and } \{D, W, Z\}$$

react with each other. Draw a graph to model these relations between the chemicals. Use this graph to find the minimum number of containers needed to ship the chemicals. Is it possible to have an allocation of the chemicals that uses the minimum number of containers and such that there are at most two chemicals in each container?

(2) The following figure shows the intersection of a major road and a small road. There are 10 traffic lanes, L_1 to L_{10}, along which vehicles approach the intersection. The directions in which vehicles along each of the lanes are allowed to negotiate the intersection and go on to a prescribed exit lane are shown. A traffic light system is installed to control movement through the intersection. The system consists of a certain number of phases. At each phase, vehicles in lanes for which the light is green may proceed safely through the intersection. What is the minimum number of phases needed for the traffic light system so that (eventually) all vehicles may proceed safely through the intersection? (We may assume that each lane is broad enough for one vehicle at a time.)

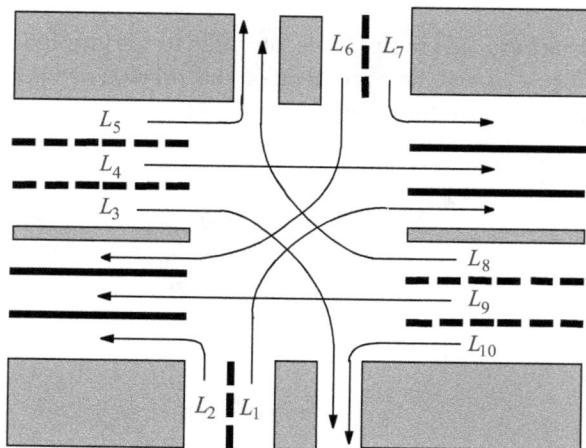

(3) A Student Council has 8 committees. Ten councilors $A, B, C, D, E, F,$ G, H, I, J are appointed to be members of the committees as shown below:

Publicity :	A, B, C, D
Recreation :	A, E, F, G
Welfare :	G, H, I, J
School Liaison :	C, J
Community :	D, E
Projects :	A, C
Secretariat :	B, F, H
Finance :	G, I

If each committee is scheduled to meet for two hours each week, what is the smallest number of two-hour sessions required to schedule all 8 committee meetings so that each of these councilors is able to attend the meetings of the committees he/she is a member of?

(4) A school is preparing a timetable for exams in 7 different subjects, labelled A to G. It is understood that if there is a pupil taking two of these subjects, their exams must be held in different timeslots. The table below shows (by crosses) the pairs of subjects which are taken by at least one pupil in common. The school wants to find the minimum number of timeslots necessary and also to allocate subjects to the timeslots accordingly. Interpreting this problem as a vertex-colouring problem, find the minimum number of timeslots needed and a suitable time allocation of the subjects.

	A	B	C	D	E	F	G
A		X	X	X		X	
B	X		X			X	X
C	X	X		X			X
D	X		X		X		
E				X		X	X
F	X	X			X		X
G		X	X		X	X	

Chapter 5

Matchings in Bipartite Graphs

5.1 Introduction

In Example 1.2.3 (see also Section 3.1), the following bipartite graph which models a job-application situation is shown:

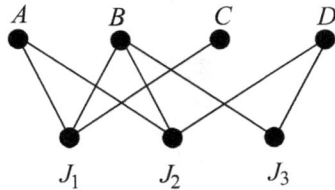

Now we ask: Is it possible to assign each applicant to a job for which he or she applies?

The answer is clearly 'no' as there are more applicants than jobs available. Indeed, the best we can do is to assign three applicants to three jobs in a way such as the following:

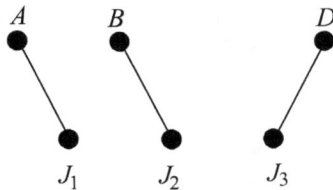

Figure 5.1

The following bipartite graph shows the acquaintance relationship between four men (M's) and five women (W's), where a vertex representing a man is adjacent to a vertex representing a woman if and only if the man is acquainted with the woman.

Again, we ask: Is it possible to marry off the four men in such a way that each man marries a woman he is acquainted with?

Though now we have more women than men, after some tries, we start to believe that this is also not possible. Indeed, the only women whom M_1, M_3 and M_4 are acquainted with are W_2 and W_4; and it is thus not possible to marry off the four men according to the condition.

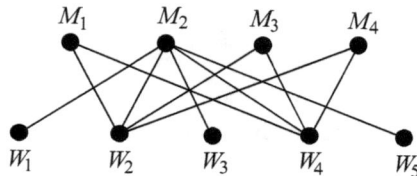

For this case, the best we can do is to marry off three of the men in a way such as the following:

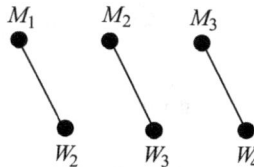

Figure 5.2

The examples above are instances respectively of the following two well-known problems.

The Assignment Problem. There are m applicants and n jobs, and each applicant is applying for a number of these jobs. Under what conditions is it possible to assign each applicant to a job for which he/she is applying?

> **The Marriage Problem.** There are m men and n women, and each man is acquainted with a certain number of the women. Under what conditions is it possible to marry off these m men in such a way that each man marries a woman he is acquainted with?

In this chapter, we shall study these problems and provide solutions to them using bipartite graphs as models.

5.2   Matchings

Figure 5.1 depicts an assignment for three applicants to three jobs, and this assignment is represented by a set of three edges in which no two edges have an end in common. The same applies to Figure 5.2. These motivate us to introduce the following:

> Let G be a graph. A non-empty set M of edges in G is called a **matching** if no two edges in M are incident with a common vertex.

Question 5.2.1.   *Consider the bipartite graph G with bipartition (X, Y) as shown in Figure 5.3.*

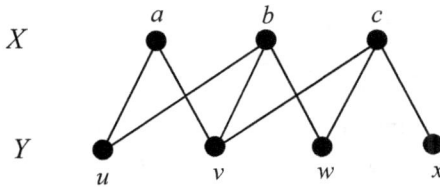

Figure 5.3

(i) Is $\{au\}$ a matching?

(ii) Is $\{av, cv\}$ a matching?

(iii) Is {au, bv, cw} a matching?

(iv) Two matchings are different if they are different as sets. Find three different matchings with three edges.

(v) Can you find in G a matching with four edges?

Question 5.2.2.

(i) Let G be a graph of order n ≥ 2, and let M be a matching in G. At most how many edges can M contain?

(ii) Find a tree of order 10 which contains a matching with 5 edges.

(iii) Find a connected graph of order 6 in which every matching can contain only one edge.

Question 5.2.3.   *Consider the bipartite graph G with bipartition (X, Y) as shown in Figure 5.4.*

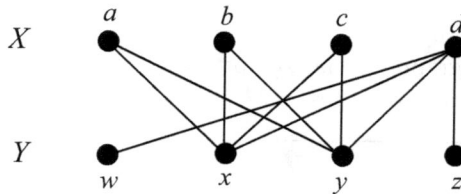

Figure 5.4

(i) Find in G a matching M with three edges, i.e., $|M| = 3$.

(ii) Does G contain a matching M with $|M| = 4 = |X|$? Why?

In the bipartite graph G of Figure 5.3, there exists a matching M such that $|M| = |X|$, i.e., every vertex in X is incident with an edge in M. This is, however, not the case in the bipartite graph of Figure 5.4.

Let G be a bipartite graph with bipartition (X, Y). A matching M in G is called a **complete matching from X to Y** if $|M| = |X|$, i.e., every vertex in X is incident with an edge in M.

Thus, in the bipartite graph of Figure 5.3, the matching

M:

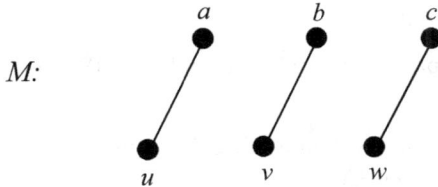

is a complete matching from X to Y (but not a complete matching from Y to X); whereas there is no complete matching from X to Y in the bipartite graph of Figure 5.4.

Let G be a graph. A matching M in G is said to be **perfect** if every vertex in G is incident with an edge in M.

In particular, if G is bipartite with bipartition (X, Y), then M is **perfect** if $|X| = |M| = |Y|$.

Question 5.2.4. *Two bipartite graphs are shown in Figure 5.5 (a) and (b).*

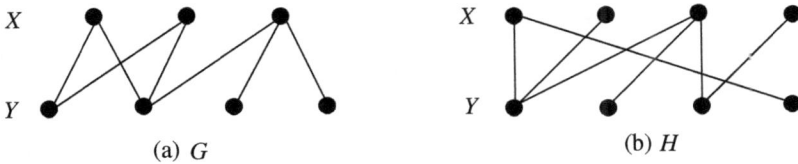

(a) G (b) H

Figure 5.5

(i) Does G have a complete matching from X to Y?

(ii) Does G have a perfect matching?

(iii) Does H have a complete matching from X to Y?

(iv) Does H have a perfect matching?

Remarks. Let G be a bipartite graph with bipartition (X, Y).

(i) If G contains a complete matching from X to Y, then $|X| \leq |Y|$; but the converse is not true.

(ii) If G contains a perfect matching, then $|X| = |Y|$; but again the converse is not true.

(iii) If G contains a complete matching M from X to Y, then M is perfect when and only when $|X| = |Y|$.

It is now clear that the assignment problem and the marriage problem as stated in Section 5.1 can be re-formulated using graph terminology as follows:

Problem. Let G be a bipartite graph with bipartition (X, Y). Under what conditions is there a complete matching from X to Y?

We shall provide a solution to this problem in the next section.

Exercise 5.2

(1) Five applicants $A_1, A_2, \cdots, A_5$ apply for five jobs $J_1, J_2, \cdots, J_5$. It is known that

    (i) J_1 is applied only by A_2,
    (ii) J_2 is applied by all except A_4,
    (iii) J_3 is applied by all except A_2,
    (iv) J_4 is applied by A_2 and A_4, and
    (v) J_5 is applied only by A_4.

    (a) Draw a bipartite graph that models the situation.
    (b) Is it possible to assign each applicant to a job for which he/she applies?

(2) In the preceding problem, suppose that the applicant A_5 changes his/her mind and applies for J_5 instead of J_2.

    (a) Draw a bipartite graph that models the situation.
    (b) Is it possible to assign each applicant to a job for which he/she applies?

(3) Five applicants apply to work in a company. There are six jobs available: $J_1, J_2, \cdots, J_6$. Applicant A is qualified for jobs J_2 and J_6; B is qualified for jobs J_1, J_3 and J_4; C is qualified for jobs J_2, J_3 and J_6; D is qualified for jobs J_1, J_2 and J_3; E is qualified for all jobs except J_4 and J_6.

    (a) Draw a bipartite graph that models the situation.
    (b) Is it possible to assign each applicant to a job for which he/she is qualified?

(4) Five men $M_1, M_2, \cdots, M_5$ and five women $W_1, W_2, \cdots, W_5$ have and only have the following acquaintance relationships between them:

    (i) each of W_1, W_2 and W_3 is acquainted with all the men,
    (ii) each of M_1 and M_5 is acquainted with all the women.

    (a) Draw a bipartite graph that models the situation.
    (b) Is it possible to marry off these five men in such a way that each man marries a woman he is acquainted with?
    (c) If M_1 insists on marrying W_1, is it possible to marry off the remaining ones in such a way that each man marries a woman he is acquainted with?

(5) Consider the following set of codewords:

$$X = \{ab, abc, cd, bcd, de\}.$$

We wish to transmit these codewords as messages. Instead of transmitting the whole codeword, we transmit a single letter which is contained in it, as its representative. Can this be done in such a way that the five codewords can be recovered uniquely from their five respective representatives?

(6) A school has vacancies for seven teachers, one for each of the subjects Chemistry, English, French, Geography, History, Mathematics and Physics. There are seven applicants for the vacancies and all are qualified to teach more than one subject. The applicants and their subjects are listed in the table below.

(a) Draw a bipartite graph to represent this situation.
(b) Determine the maximum number of (suitably qualified) teachers the school can employ.

Applicants	Subjects qualified
Miss Lim	Mathematics, Physics
Miss Wong	Chemistry, English, Mathematics
Miss Tay	Chemistry, French, History, Physics
Mr. Tan	English, French, History, Physics
Mr. Lee	Chemistry, Mathematics
Mr. Ng	Mathematics, Physics
Mr. Peng	English, Geography, History

(7) Consider the following bipartite graph G with bipartition (X, Y):

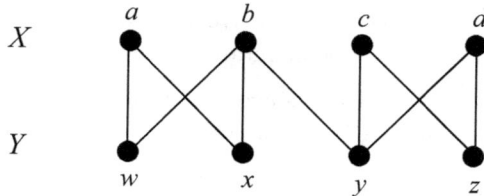

(i) Is $\{bx, cy\}$ a matching?
(ii) Is $\{ax, by, cy\}$ a matching?
(iii) Is $\{ax, by, cz\}$ a matching?
(iv) Is $\{ax, bw, cz, dy\}$ a matching? a perfect matching?

(v) Is there any perfect matching that contains the edge 'by'?

(vi) Find the number of perfect matchings in G.

(8) For $n \geq 3$, find the number of perfect matchings of the cycle C_n.

(9) For $n \geq 1$, find the number of perfect matchings of the graph $K(n,n)$.

(10) (+) For $n \geq 2$, find the number of perfect matchings of the graph K_n.

(11) Consider the following bipartite graph G with bipartition (X, Y):

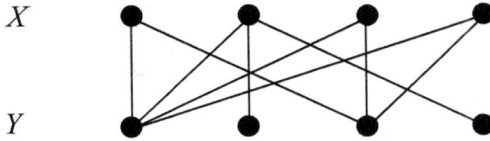

(i) Does G contain a complete matching from X to Y?

(ii) Does G contain a perfect matching?

(iii) Find two matchings M and M' in G with $|M| = |M'| = 3$.

(iv) How many matchings M are there in G with $|M| = 3$?

(12) (+) Consider the following 6×6 grid-board whose upper left and lower right corner squares are removed. You are given 17 dominoes, each covering exactly two adjacent squares (squares that have an edge in common) of the board. Can you use them to cover the 34 squares in the board?

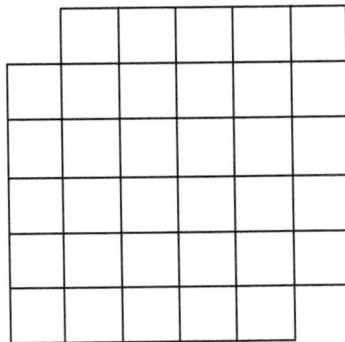

(13) (+) Prove that the following 3-regular graph (triple flyswat) does not have a perfect matching, but does have a matching with seven edges.

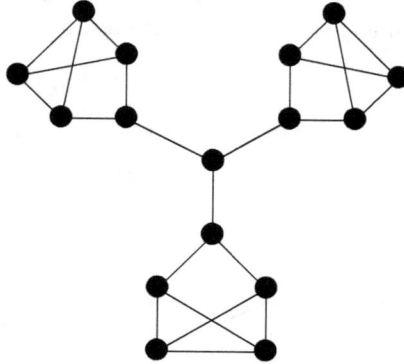

(14) Let T be a tree of order $n \geq 2$ and M, a matching in T.

  (i) What is the largest possible value for $|M|$? Construct one such T which contains one such M having its $|M|$ attaining this largest value.

  (ii) What is the least possible value for $|M|$? Construct one such T which contains one such M having its $|M|$ attaining this least value.

(15) (+) The twenty members of a local tennis club have scheduled exactly 14 two-person games among themselves, with each member playing in at least one game. Prove that within this schedule there must be a set of 6 games with twelve distinct players. (USAMO, 1989)

(16) (+) Generalize the result in the preceding problem by replacing 'twenty members' by 'n members', and '14 games' by 'm games'.

5.3 Hall's theorem

Before providing a solution to the problem stated at the end of Section 5.2, let us first introduce the following notation.

Let G be a graph. For a set S of vertices in G, let $N(S)$ denote the set of vertices in G which are adjacent to some vertex in S; i.e.,

$$N(S) = \{v \in V(G)|v \text{ is adjacent to some vertex in } S\}.$$

It is clear that

$$N(S) = \bigcup_{a \in S} N(a).$$

Question 5.3.1. *Let G be the bipartite graph with bipartition (X, Y) as shown in Figure 5.6.*

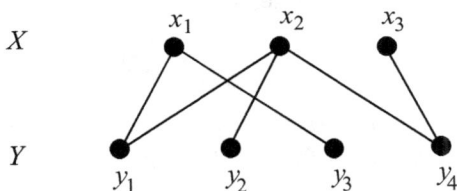

Figure 5.6

(i) Complete the following table:

S	$N(S)$
$\{x_1\}$	$\{y_1, y_3\}$
$\{x_2\}$	
$\{x_3\}$	
$\{x_1, x_2\}$	
$\{x_1, x_3\}$	$\{y_1, y_3, y_4\}$
$\{x_2, x_3\}$	
X	

(ii) Is there a complete matching from X to Y?

Question 5.3.2. *Consider the bipartite graph G with bipartition (X, Y) as shown in Figure 5.7.*

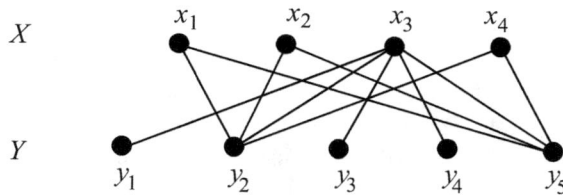

Figure 5.7

(i) Let $S = \{x_1, x_2, x_4\}$. Find $N(S)$.

(ii) Is there a complete matching from X to Y in G? Why?

We are now in a position to establish the following classic result on matchings in bipartite graphs due to the English algebraist Philip Hall (1904 – 1982) in 1935.

Theorem 5.1. *Let G be a bipartite graph with bipartition (X, Y). Then G contains a complete matching from X to Y if and only if $|S| \leq |N(S)|$ for every subset S of X.*

Proof. [Necessity] Suppose on the contrary that there exists $S \subseteq X$ such that $|S| > |N(S)|$. Then, for any matching M of G, $|V(M) \cap S| \leq |N(S)| < |S|$, where $V(M)$ is the set of vertices in G incident with edges in M, implying that M is not a complete matching from X to Y. Thus, the necessity holds.

[Sufficiency] Assume that in G

$(\#)$ $|S| \leq |N(S)|$ for all $S \subseteq X$.

We shall show that G contains a complete matching from X to Y by induction on $|X|$. The above statement is obviously true if $|X| = 1$. Assume that it is true when $|X| \leq k - 1$. Consider now that $|X| = k$, where $k \geq 2$.

Case (1). $|S| + 1 \leq |N(S)|$ for all $S \subseteq X$ with $S \neq \emptyset$ and $S \neq X$.

Let $x \in X$. Then there exists $y \in Y$ such that $xy \in E(G)$. Let $G' = G - \{x, y\}$. Clearly, G' satisfies $(\#)$ for all $S \subseteq X - \{x\}$. Thus, by the induction hypothesis, G' contains a complete matching M' from $X - \{x\}$ to $Y - \{y\}$. It follows that $M' \cup \{xy\}$ is a complete matching from X to Y.

Case (2). $|S_0| = |N(S_0)|$ for some $S_0 \subseteq X$ with $S_0 \neq \emptyset$ and $S_0 \neq X$.

(i) Consider the subgraph $G_0 = [S_0 \cup N(S_0)]$ of G. Observe that G_0 is also bipartite, and satisfies (#) (with X replaced by S_0). By the induction hypothesis, G_0 contains a complete matching M_0 from S_0 to $N(S_0)$.

(ii) Consider $G' = G - (S_0 \cup N(S_0))$.

Claim: G' satisfies (#) (with X replaced by $X \backslash S_0$).

If not, then there exists $S_1 \subseteq X \backslash S_0$ such that $|S_1| > |N(S_1)|$ in G'. But then $|S_0 \cup S_1| > |N(S_0 \cup S_1)|$ in G, a contradiction.

Thus, the above claim holds, and by the induction hypothesis, G' contains a complete matching M' from $X \backslash S_0$ to $Y \backslash N(S_0)$.

Now, combining (i) and (ii), we obtain a matching M, namely $M_0 \cup M'$, which is a complete matching from X to Y in G. $\square$

As an immediate consequence of Theorem 5.1, we have:

Corollary 5.2. *Let G be a bipartite graph with bipartition (X, Y) such that $|X| = |Y|$. Then G has a perfect matching if and only if $|S| \leq |N(S)|$ for every subset S of X.* $\square$

Example 5.3.1. *In Question 5.3.1, the bipartite graph G with bipartition (X, Y) as depicted in Figure 5.6 has a complete matching from X to Y by Theorem 5.1 since, as shown in the table therein, $|S| \leq |N(S)|$ for every subset S of X.*

On the other hand, in Question 5.3.2, the bipartite graph G with bipartition (X, Y) as depicted in Figure 5.7 has no complete matching from X to Y by Theorem 5.1 as there exists a subset S of X, namely, $S = \{x_1, x_2, x_4\}$, such that $|S| > |N(S)|$.

A special family of bipartite graphs which always contain perfect matchings is given in the following.

Example 5.3.2. *Figure 5.8 shows a 3-regular bipartite graph, and it is easy to see from the diagram that it contains a perfect matching. Indeed, every 3-regular bipartite graph always contains a perfect matching.*

X

Y

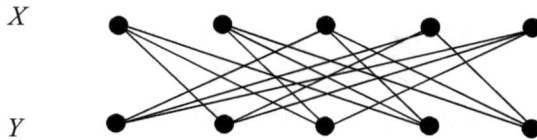

Figure 5.8

Let us now argue why this is the case.

Proof. Let G be a 3-regular bipartite graph with bipartition (X, Y).

We shall apply Theorem 5.1, and our target is to show that

$$|S| \leq |N(S)|$$

for any subset S of X.

Let $S \subseteq X$ be given.

Firstly, we ask: (1) How many edges in G are incident with some vertex in S?

Since G is 3-regular, the answer is $3|S|$.

Next, we ask: (2) How many edges in G are incident with some vertex in $N(S)$?

Again, by the same reason, the answer is $3|N(S)|$.

Since an edge incident with a vertex in S must be incident with a vertex in $N(S)$, any edge counted in (1) must be counted in (2). Thus, we have (see Problem 5 of Exercise 5.3):

$$3|S| \leq 3|N(S)|,$$

that is, $|S| \leq |N(S)|$, as required.

Thus, by Theorem 5.1, G contains a complete matching from X to Y.

Since G is a regular bipartite graph with bipartition (X, Y), we must have $|X| = |Y|$ (see Problem 4 of Exercise 3.1). We therefore conclude that the above complete matching from X to Y is a perfect matching. This completes the proof.   □

In general, we have (see Problem 6 of Exercise 5.3):

Corollary 5.3. *Every k-regular bipartite graph, where $k \geq 1$, always contains a perfect matching.*   □

Exercise 5.3

(1) Let G be the bipartite graph you constructed in Problem 1 of Exercise 5.2. Let $S = \{A_1, A_3, A_5\}$. Find $N(S)$. Is $|N(S)| < |S|$? What conclusion can you draw from Theorem 5.1?

(2) Let G be the bipartite graph with bipartition (X, Y) as shown in Problem 7 of Exercise 5.2.

(i) Complete the following table:

S	$N(S)$
$\{a\}$	
$\{b\}$	
$\{c\}$	
$\{d\}$	
$\{a, b\}$	
$\{a, c\}$	
$\{a, d\}$	
$\{b, c\}$	
$\{b, d\}$	
$\{c, d\}$	
$\{a, b, c\}$	
$\{a, b, d\}$	
$\{a, c, d\}$	
$\{b, c, d\}$	
X	

(ii) Is it true that $|S| \leq |N(S)|$ for all $S \subseteq X$?

(iii) What conclusion can you draw from Corollary 5.2?

(3) Let G be the bipartite graph with bipartition (X, Y) as shown in Problem 11 of Exercise 5.2.

(i) Find a subset S of X such that $|S| > |N(S)|$.

(ii) What conclusion can you draw from Theorem 5.1?

(4) Consider the following bipartite graph G with bipartition (X, Y):

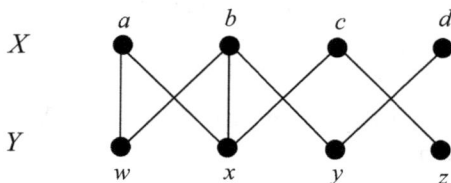

(i) Let $S = \{a, b\}$. Find $N(S)$.

(ii) Let E_1 be the set of edges in G incident with some vertex in S. Find E_1.

(iii) Let E_2 be the set of edges in G incident with some vertex in $N(S)$. Find E_2.

(iv) Is $E_1 \subseteq E_2$?

(5) Let G be a bipartite graph with bipartition (X, Y). For $S \subseteq X$, let E_1 be the set of edges in G incident with some vertex in S, and let E_2 be the set of edges in G incident with some vertex in $N(S)$. Is it true in general that $E_1 \subseteq E_2$? Why?

(6) $(+)$ Let G be a bipartite graph with bipartition (X, Y). Assume that there exists a positive integer k such that $d(y) \leq k \leq d(x)$ for each vertex y in Y and each vertex x in X.

Let $S \subseteq X$, and denote by E_1 the set of edges in G incident with some vertex in S, and by E_2 the set of edges in G incident with some vertex in $N(S)$ (see the preceding problem).

(i) Show that $k|S| \leq |E_1| \leq |E_2| \leq k|N(S)|$.

(ii) Deduce from Theorem 5.1 that G has a complete matching from X to Y.

(iii) Deduce from (ii) that every k-regular bipartite graph with $k \geq 1$ has a perfect matching.

(7) $(+)$ Let G be a bipartite graph with bipartition (X, Y). Let $\Delta(G) = k \geq 1$ and $X^* = \{x \in X \mid d(x) = k\}$. Assume that X^* is not empty. Determine whether the following statement $(\#)$ is true and justify your answer.

$(\#)$ G contains a matching M such that every vertex in X^* is incident with an edge in M.

(8) (+) Let G be a bipartite graph with bipartition (X, Y). Prove that G contains a complete matching from X to Y if and only if

$$|X \backslash N(T)| \leq |Y \backslash T|$$

for all $T \subseteq Y$.

(9) Let G be a bipartite graph. Prove that G contains a perfect matching if and only if $|S| \leq |N(S)|$ for all $S \subseteq V(G)$.

(10) (+) Let G be a connected bipartite graph with bipartition (X, Y), where $|X| \geq 2$ and $|Y| \geq 2$. Prove that the following statements are equivalent:

   (i) Each edge of G is contained in a perfect matching of G.
   (ii) $|X| = |Y|$ and $|S| < |N(S)|$ for all $S \subset X$ with $S \neq \emptyset$.
   (iii) $G - \{x, y\}$ has a perfect matching for any $x \in X$ and $y \in Y$.

(11) (+) Let G be a bipartite graph with bipartition (X, Y) such that $|X| - |Y| = p \geq 1$. Form a larger bipartite graph G^* with bipartition $(X, Y \cup Y^*)$, where $|Y^*| = p$, such that
   (1) G^* contains G as an induced subgraph, and
   (2) every vertex in Y^* is adjacent to every vertex in X.
Prove that G^* has a perfect matching if and only if G has a matching with $|Y|$ edges.

(12) (+) Let G be a bipartite graph with bipartition (X, Y), and let k be an integer such that $1 \leq k \leq |X|$. Show that G contains a matching M with $|M| = k$ if and only if

$$|S| \leq |N(S)| + |X| - k$$

for all $S \subseteq X$.

(13) (+) Let G be a bipartite graph with bipartition (X, Y). Assume that $d(x) \geq 6$ and $d(y) \leq 8$ for all $x \in X$ and $y \in Y$. Prove that G contains a matching M with $|M| \geq \frac{3}{4}|X|$.

(14) (+) Let G be a bipartite graph with bipartition (X, Y). For $S \subseteq X$, the **deficiency** $\rho(S)$ of S is defined as

$$\rho(S) = |S| - |N(S)|.$$

Assume that $d(x) \geq 3$ and $d(y) \leq 4$ for all $x \in X$ and $y \in Y$. Show that

$$|X| \geq 4\rho(S),$$

for all $S \subseteq X$.

(15) (+) Let k and n be positive integers with $k \leq n$. A $k \times n$ matrix with entries from $\{1, 2, \cdots, n\}$ is called a **Latin rectangle** if each 'i' in $\{1, 2, \cdots, n\}$ appears exactly once in each row and at most once in each column. A $k \times n$ Latin rectangle is called a **Latin square** of order n if $k = n$.

Consider the 3×5 Latin rectangle $L = \begin{pmatrix} 1\ 2\ 3\ 4\ 5 \\ 5\ 1\ 2\ 3\ 4 \\ 4\ 5\ 1\ 2\ 3 \end{pmatrix}$.

Define a bipartite graph G with bipartition (X, Y) associated with L as follows:

(i) $X = \{1, 2, 3, 4, 5\}$,

(ii) $Y = \{C_1, C_2, C_3, C_4, C_5\}$, where C_i is the ith column of L,

(iii) 'i' in X is adjacent to 'C_j' in Y if and only if 'i' does not appear in 'C_j'.

(a) Draw the diagram of G.

(b) What is the degree of each vertex in X? Why?

(c) What is the degree of each vertex in Y? Why?

(d) Is G 2-regular?

(e) Does G contain a perfect matching? Why?

(f) Display a perfect matching in G if your answer to (e) is 'yes'.

(g) Use the perfect matching obtained in (f) to append a new row to L to form a 4×5 Latin rectangle L'.

(h) Expand L' to form a Latin square of order 5.

(16) (+) Consider the 2×6 Latin rectangle $L = \begin{pmatrix} 1\ 2\ 3\ 4\ 5\ 6 \\ 3\ 6\ 4\ 5\ 2\ 1 \end{pmatrix}$.

Define, likewise, the bipartite graph G with bipartition (X, Y) associated with L as shown in the preceding problem by replacing X by $\{1, 2, 3, 4, 5, 6\}$, and Y by $\{C_1, C_2, C_3, C_4, C_5, C_6\}$, where C_i is the ith column of L.

(a) Draw the diagram of G.

(b) What is the degree of each vertex in X? Why?

(c) What is the degree of each vertex in Y? Why?

(d) Is G 4-regular?

(e) Does G contain a perfect matching? Why?

(f) Display a perfect matching in G if your answer to (e) is 'yes'.

(g) Use the perfect matching obtained in (f) to append a new row to L to form a 3×6 Latin rectangle L'.

5.4 System of distinct representatives

Let us begin with the following example.

Example 5.4.1. *At a junior college, there are four clubs with their executives as shown below:*

(B) Biology club: {T, V},
(C) Chemistry club: {U, V, Z},
(M) Mathematics club: {U, V},
(P) Physics club: {R, Z}.

The college wishes to form a student council which includes one representative from each club with the condition that the representative of a club must be one of its executives and different clubs must have different representatives. Can this be done?

Yes! One possible solution is, for instance, the following:

(B) is represented by T,
(C) is represented by U,
(M) is represented by V,
(P) is represented by Z.

and

This example motivates us to introduce the following notion.

Let W be a non-empty set, and $S_1, S_2, \cdots, S_m$ be non-empty finite subsets (*but not necessarily distinct*) of W. A **system of distinct representatives (SDR)** for the family $(S_1, S_2, \cdots, S_m)$ is an m-tuple $(a_1, a_2, \cdots, a_m)$ such that $a_i \in S_i$ for each $i = 1, 2, \cdots, m$, and $a_i \neq a_j$ whenever $i \neq j$.

Question 5.4.1. *Let W be the set of natural numbers. Consider the following subsets of W:*

$$S_1 = \{1, 2\}, S_2 = \{2, 5\}, S_3 = \{4\}, S_4 = \{3, 5\} \text{ and } S_5 = \{3, 5\}.$$

Does the family $(S_1, S_2, \cdots, S_5)$ have an SDR?

Remark. In the above question, S_4 and S_5 are the same, which is fine.

Question 5.4.2.   *Let W be the set of natural numbers.   Consider the following subsets of W:*

$$S_1 = \{1\}, S_2 = \{2,3\}, S_3 = \{1,4,5\}, S_4 = \{1,2\} \text{ and } S_5 = \{1,3\}.$$

Does the family $(S_1, S_2, \cdots, S_5)$ have an SDR?

The family $(S_1, S_2, \cdots, S_5)$ in Question 5.4.2 does not have an SDR. Why?

Look at the four sets S_1, S_2, S_4 and S_5 in the family. Notice that

$$S_1 \cup S_2 \cup S_4 \cup S_5 = \{1,2,3\}.$$

We see that there are more sets (4 sets) than members (3 only), and that is why the family does not have an SDR.

In general, given a family of sets $(S_1, S_2, \cdots, S_m)$, it is clear that if there exist k of them whose union has less than k members, then the family does not have any SDR. That is, if there exists some $I \subseteq \{1, 2, \cdots, m\}$ such that

$$\left| \bigcup_{i \in I} S_i \right| < |I|,$$

then the family does not have any SDR.

Is the converse true? That is, if

$$\left| \bigcup_{i \in I} S_i \right| \geq |I|,$$

for all subsets I of $\{1, 2, \cdots, m\}$, is it true that the family $(S_1, S_2, \cdots, S_m)$ always has an SDR?

We shall now examine this using Example 5.4.1. Firstly, we note that for any number of clubs from $\{(B), (C), (M), (P)\}$, the number of executives in these clubs is always greater than or equal to the number of these clubs (for instance, clubs (B) and (C) altogether have four executives, namely, T, U, V and Z).

Let $X = \{(B), (C), (M), (P)\}$, i.e., the set of four clubs, and

$$Y = \{R, T, U, V, Z\},$$

i.e., the set of executives in these clubs.

We construct a bipartite graph G with bipartition (X, Y) by defining the following adjacency: a vertex (club) in X is adjacent to a vertex (person) in Y if and only if that person is an executive of the club.

Following this, the graph G is depicted in Figure 5.9.

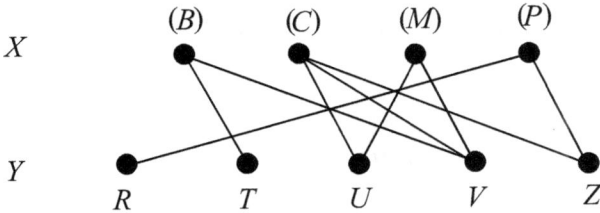

Figure 5.9

Our next target is to show that

$$|S| \leq |N(S)|$$

for every subset S of X.

Thus, let S be a subset of X, say $S = \{(B), (M)\}$. We ask: What is $N(S)$?

By the definition of the adjacency, $N(S)$ consists of those executives of (B) or (M); that is, $N(S) = \{T, U, V\}$. Clearly, we have

$$|S| \leq |N(S)|.$$

It can similarly be verified that the above inequality holds for all subsets S of X.

Fine! We can now apply Theorem 5.1 to obtain a complete matching from X to Y, for instance,

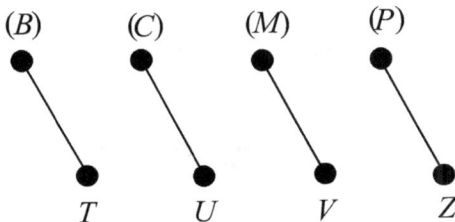

What does this matching mean to us? Well, it simply means that (T, U, V, Z) is an SDR of the family $((B), (C), (M), (P))$.

We now end this chapter by establishing the following general result on SDR.

Theorem 5.4. *The family $(S_1, S_2, \cdots, S_m)$ of non-empty finite subsets of a set W has an SDR if and only if*

$$\left| \bigcup_{i \in I} S_i \right| \geq |I|$$

for all subsets I of $\{1, 2, \cdots, m\}$.

Proof. [Necessity] Let $(a_1, a_2, \cdots, a_m)$ be an SDR of $(S_1, S_2, \cdots, S_m)$, and let $I \subseteq \{1, 2, \cdots, m\}$. Then

$$\{a_i | i \in I\} \subseteq \bigcup_{i \in I} S_i,$$

and so

$$\left| \bigcup_{i \in I} S_i \right| \geq |\{a_i | i \in I\}| = |I|.$$

[Sufficiency] Let $X = \{1, 2, \cdots, m\}$ and $Y = \bigcup_{i=1}^{m} S_i$. Construct a bipartite graph G with bipartition (X, Y) such that $i(\in X)$ is adjacent to $y(\in Y)$ if and only if $y \in S_i$.

Let $I \subseteq X$. Then in G

$$N(I) = \bigcup_{i \in I} S_i$$

and so by assumption,

$$|N(I)| = \left| \bigcup_{i \in I} S_i \right| \geq |I|.$$

By Theorem 5.1, G contains a complete matching from X to Y:

$$1y_1, \ 2y_2, \ \cdots, \ my_m,$$

where $y_r \neq y_s$ whenever $r \neq s$. Clearly, $(y_1, y_2, \cdots, y_m)$ is an SDR of $(S_1, S_2, \cdots, S_m)$. $\qquad \square$

Exercise 5.4

(1) Find an SDR for each of the following families of sets.

    (i) $(\{1,2\}, \{2,5\}, \{4\}, \{3,5\})$;

    (ii) $(\{1,3\}, \{1,3\}, \{2\}, \{1,4\}, \{4,5\})$;

    (iii) $(\{5\}, \{1,6\}, \{2,3\}, \{1,4\}, \{3\}, \{1,4\})$.

(2) For each of the following families of sets, determine whether it has an SDR. Justify your answers.

    (i) $(\{1\}, \{2,3\}, \{1,2\}, \{1,3\}, \{1,4,5\})$;

    (ii) $(\{1,2\}, \{2,3\}, \{4,5\}, \{4,5\})$;

    (iii) $(\{1,2\}, \{2,3\}, \{3,4\}, \{4,5\}, \{5,1\})$.

(3) There are four clubs in a school with their committee members as shown below:

Club (A): $\{a, b\}$,

Club (B): $\{a, c, e\}$,

Club (C): $\{b, c\}$,

Club (D): $\{b, d\}$.

Let $X = \{A, B, C, D\}$ and $Y = \{a, b, c, d, e\}$. We construct a bipartite graph G with bipartition (X, Y) as follows: A vertex (club) in X is adjacent to a vertex (person) in Y if and only if that person is a committee member of the club.

    (i) Draw the graph G.

    (ii) Let $S = \{A\}$. Find $N(S)$ in G.

    (iii) Let $S = \{A, B\}$. Find $N(S)$ in G.

    (iv) Let $S = \{A, B, C\}$. Find $N(S)$ in G.

    (v) Does there exist a subset S of X such that $|S| > |N(S)|$?

    (vi) Does there exist a complete matching from X to Y?

    (vii) Display a complete matching M from X to Y if your answer to (vi) is 'yes'.

  (viii) Provide an SDR for the family (A, B, C, D) from M.

(4) Let $S_1 = \{b, c\}$, $S_2 = \{a\}$, $S_3 = \{a, b\}$ and $S_4 = \{c, d\}$. Verify that the family (S_1, S_2, S_3, S_4) satisfies the condition stated in Theorem 5.4, and thus conclude that the family has an SDR. Provide also one such SDR.

(5) Six teachers A, B, C, D, E and F are members of five committees. The memberships of the committees are $\{A, B, C\}$, $\{D, E, F\}$, $\{A, D, E, F\}$, $\{A, C, E, F\}$, and $\{A, B, F\}$. The activities of each com-

mittee are to be reviewed by a teacher who is not on the committee, and different committees are to be reviewed by different teachers. Can five distinct teachers be selected? If 'yes', show one such assignment.

(6) Show that each of the following families of sets has **no** SDR by Theorem 5.4.

    (i) $(\{1,2\},\{1\},\{3,4\},\{2\})$;

    (ii) $(\{1\},\{2,3\},\{1,4,5\},\{1,2\},\{1,3\})$;

    (iii) $(\{2,3\},\{2,3,4,5,6\},\{3,4\},\{4,5\},\{2,5\},\{2,4\})$.

(7) For $n \geq 2$, let $S_1 = \{1\}$, $S_2 = \{1,2\}$ and for each $i = 3, \cdots, n$, let $S_i = \{1,2,\cdots,i\}$.

    (i) Show that the family $(S_1, S_2, \cdots, S_n)$ has an SDR.

    (ii) How many different SDR's does $(S_1, S_2, \cdots, S_n)$ have?

(8) For $n \geq 2$ and for each $i = 1,2,\cdots,n-1$, let $S_i = \{i, i+1\}$, and $S_n = \{n, 1\}$.

    (i) Show that the family $(S_1, S_2, \cdots, S_n)$ has an SDR.

    (ii) How many different SDR's does $(S_1, S_2, \cdots, S_n)$ have?

(9) (+) Let $S_1 = \{1, a\}$, $S_2 = \{1, 2a - 1\}$, $S_3 = \{2, 4 - a\}$ and $S_4 = \{2, a + 1\}$, where $a \in \{1,2,3,4\}$. Find all possible values of 'a' for which the family (S_1, S_2, S_3, S_4) has an SDR.

(10) (+) There are 12 clubs at a junior college. It is known that each club has at least 3 members and no student is a member of four or more clubs. Prove that this family of 12 clubs has an SDR.

Chapter 6

Eulerian Multigraphs and Hamiltonian Graphs

6.1 Eulerian multigraphs

We introduced the Königsberg bridge problem at the beginning of this book and, in Section 1.3, we mentioned how Euler generalized it to a much more general problem. As promised then, we will now discuss Euler's solution in this chapter.

Let us begin with the following example to recall some relevant concepts.

Example 6.1.1. *Consider the multigraph G of Figure 6.1(a) and the following walk in G,*

$$W: \quad xe_1we_4ye_5we_3xe_7ze_8ye_9ze_6we_2x.$$

The ordering of the edges following the walk is shown in Figure 6.1(b).

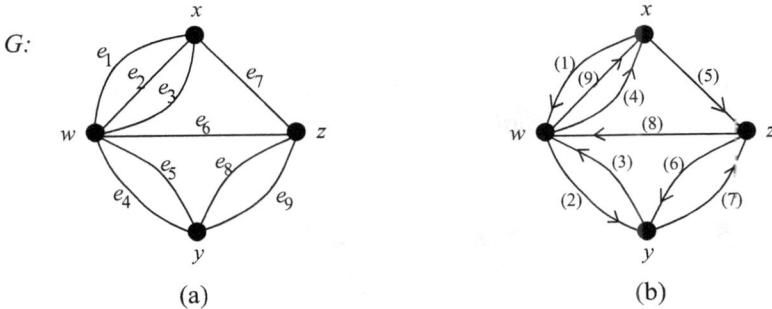

(a)   (b)

Figure 6.1

157

(i) Is there any edge repeated in W?

No! No edge in W is repeated. Thus, W is a trail.

(ii) Is W a closed trail?

Yes! W begins and terminates at x. Thus, W is a closed trail; that is, W is a circuit.

(iii) Does W include all the edges in G?

Yes! All the nine edges in G are contained in W.

The features of the walk W in Example 6.1.1 remind us of the following problem studied by Euler which was introduced in Section 1.3:

Let G be a multigraph. Assume that, starting with an arbitrary vertex in G, we can have a walk which passes through each edge once and only once, and then be able to terminate at the starting vertex. What can be said about G?

Let G be a connected multigraph. A circuit W in G is called an **Euler circuit** if W contains all the edges in G. The multigraph G is called an **Eulerian multigraph** if G possesses an Euler circuit.

Thus, in Example 6.1.1, the walk W is an Euler circuit of G, and so G is an Eulerian multigraph.

Remark. In this and the next sections, we shall **not** confine ourselves to 'simple graphs'.

Question 6.1.1. *Show that each of the following multigraphs is Eulerian by exhibiting an Euler circuit.*

Is there an odd vertex in each multigraph?

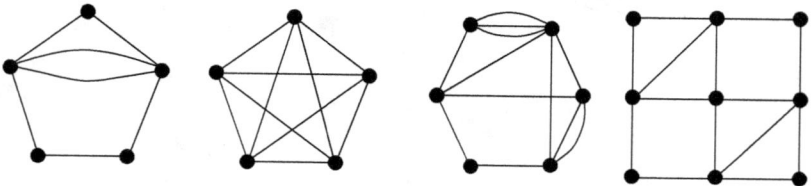

Question 6.1.2.   *Are the following multigraphs Eulerian? Are there any odd vertices in each multigraph?*

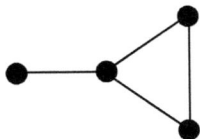

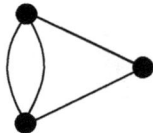

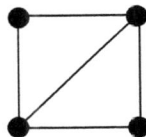

Exercise 6.1

(1) Consider the following multigraph G of order 5.

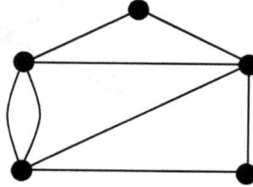

    (i) Find $e(G)$.
   (ii) Find in G a circuit with 2 edges; with 3 edges; with 4 edges.
  (iii) Find in G a circuit with 5 edges that is not a cycle.
  (iv) Find in G a circuit with 6 edges.
   (v) If W is an Euler circuit in G, exactly how many edges are contained
       in W?
  (vi) Does G contain an Euler circuit? Show one if there is.

(2) Five multigraphs are depicted below. Show that each of them is Eulerian by exhibiting an Euler circuit.

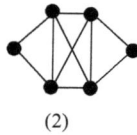

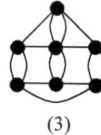

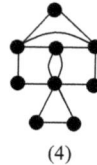

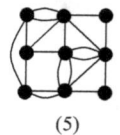

(1)   (2)   (3)   (4)   (5)

6.2   Characterization of Eulerian multigraphs

Let us recall some observations found in the discussion of Questions 6.1.1 and 6.1.2. We see that all the multigraphs in Question 6.1.1 are Eulerian and at the same time they contain no odd vertices. On the other hand, all multigraphs in Question 6.1.2 are not Eulerian and they contain some odd vertices.

A natural question is: *Does the fact that a given multigraph G is Eulerian or not depend on the non-existence of odd vertices in G?*

Let us study this question. Suppose that G is an Eulerian multigraph. Then, by definition, G possesses an Euler circuit $v_1 e_1 v_2 e_2 \cdots v_m e_m v_{m+1}$, denoted by W, which passes through each edge of G once and exactly once. Thus $e_1, e_2, \cdots, e_m$ are the distinct edges in G, and each edge in G is one of the e_i's. Note that the vertices $v_1, v_2, \cdots, v_{m+1}$ need not be distinct (indeed, $v_1 = v_{m+1}$).

We now claim that *each vertex in G must be even*. To see this, let v be an arbitrary vertex in G. Assume first that v is not the initial vertex in the Euler circuit W (hence v is also not the terminal vertex in W). Then each time we traverse W to visit v, there must be two edges in W, say e_i and e_{i+1}, such that the former one is for us to reach v and the latter one for us to leave v. Since all the edges incident with v are contained in the walk W, the number of edges incident with v is thus even, which means that v is an even vertex. Assume now that v is the initial vertex (and so the terminal vertex also) in W. That is, $v = v_1 = v_{m+1}$. For the first move, there is an edge (i.e., e_1) for us to leave v; for the last move, there is an edge (i.e., e_m) for us to return to v; and besides these, each time (if ever) we visit v there must be two edges in W, one for entering v and one for leaving v. Thus, again, the number of edges incident with v is even; that is, v is an even vertex.

What we discussed above was due to Euler [E], and from it, Euler drew the following conclusion:

Result (#): If G is an Eulerian multigraph, then each vertex in G is even.

We now re-visit the Königsberg bridge problem. The multigraph that models the situation is re-depicted in Figure 6.2.

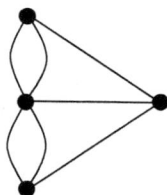

Figure 6.2

Notice that no vertex in it is even. It follows immediately from Result (#) that the answer to the Königsberg bridge problem is 'no'. Likewise, all the multigraphs in Question 6.1.2 contain some odd vertices, and accordingly, they are all not Eulerian.

Euler's discovery says that if G is Eulerian, then every vertex in G is even. Is the converse true? That is, if G is a connected multigraph in which every vertex is even, must G be Eulerian?

Euler [E] thought that the answer was 'yes', but did not provide a proof.

Unaware of Euler's work, Carl Hierholzer, a young German mathematician, published his work [H] in 1873 which contains not only a proof of Euler's discovery, but also a proof of its converse (see Problem 7 of Exercise 6.2). Thus we have now the following characterization of Eulerian multigraphs.

Theorem 6.1.   *Let G be a connected multigraph. Then G is Eulerian if and only if every vertex in G is even.*   □

Remark. To determine whether or not a given connected multigraph G is Eulerian according to the definition by trying our luck searching for an Euler circuit in G is by no means a simple task, especially when G contains a large number of edges. On the other hand, checking whether or not a vertex is even is really just a small matter. Thus, Theorem 6.1 reduces the amount of work required to determine if G is Eulerian.

Observe that the multigraph on the cover of this book has only even vertices, and thus it is Eulerian. If we start from the 'black vertex' of degree 4 and traverse the edges according to the order of the colours of the spectrum (i.e. Violet, Indigo, Blue, Green, Yellow, Orange, Red), we have an Euler circuit.

Question 6.2.1.   *(i) Is every cycle Eulerian?*

(ii) Can any tree be Eulerian?

(iii) Which K_n, $n \geq 2$, are Eulerian?

(iv) Which $K(p, q)$, $p \geq q \geq 2$, are Eulerian?

Example 6.2.1.   *Though none of the multigraphs in Question 6.1.2 is Eulerian (that is, none of them contains Euler circuits), careful checking shows that each of them does possess an **open** walk which passes through each of its edges once and exactly once as depicted below:*

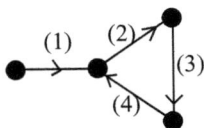

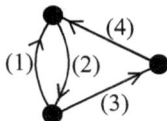

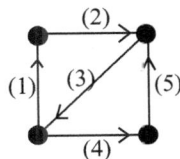

Let G be a connected multigraph. A trail W in G is called an **Euler trail** if W contains all the edges in G. The multigraph G is called a **semi-Eulerian multigraph** if G possesses an open Euler trail.

Thus, all multigraphs in Question 6.1.2 are semi-Eulerian.

Question 6.2.2.   *Is the following multigraph semi-Eulerian? If it is so, find an open Euler trail in it.*

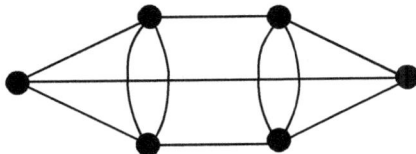

We are sure that you are able to find an open Euler trail in the multigraph G shown in Question 6.2.2. Thus G is semi-Eulerian. Now, G certainly contains odd vertices; but how many odd vertices does G contain? Which vertices are the initial vertex and the terminal vertex in your open Euler trail of G?

As shown in Figure 6.3 (a), G has exactly two odd vertices, namely, u and v. The open Euler trail of G shown in Figure 6.3(b) has u as its initial vertex and v as its terminal vertex.

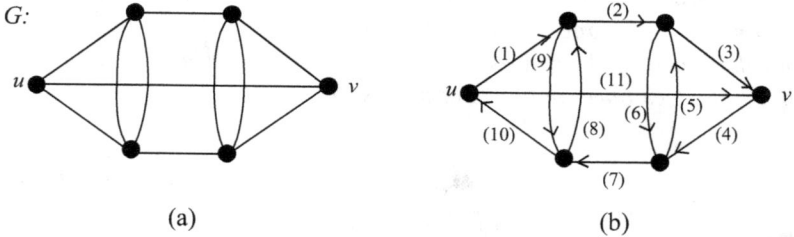

Figure 6.3

Indeed, in general, we have (see Problem 8 of Exercise 6.2):

Theorem 6.2.   *Let G be a connected multigraph.   Then G is semi-Eulerian if and only if G has exactly two odd vertices.   Moreover, if G is semi-Eulerian, then the two odd vertices in G are the initial and terminal vertices of any Euler trail in G.*   □

Question 6.2.3.   *In each of the following multigraphs, find the number of odd vertices. Which ones are semi-Eulerian?*

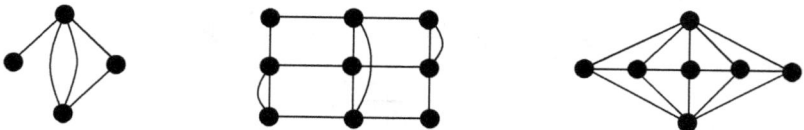

Question 6.2.4.   *The following multigraph G is semi-Eulerian as it contains exactly two odd vertices, namely, v and x.*

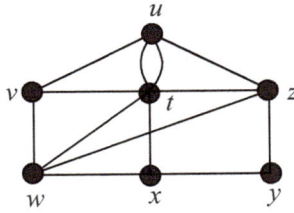

(i) *Form a new multigraph G^* by adding to G a new edge joining v and x. Is G^* Eulerian? Why?*

(ii) *Find an Euler circuit W in G^* (note that the new edge vx is contained in W).*

(iii) *Delete vx (or xv) from W. Can you find an open Euler trail of G from the resulting sequence of edges?*

An Extension – The Chinese Postman Problem (CPP)

Figure 6.4 shows a map of streets in an area where P is a local post office. Leaving from P, a postman needs to deliver mail along every street in this area, and then return to P. How should his route be planned so that the postman walks as little as possible?

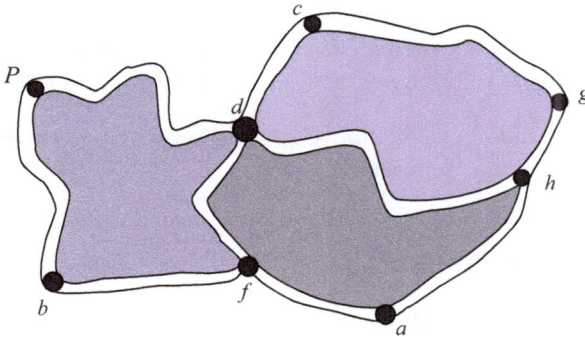

Figure 6.4

The graph G that models the above street section is shown in Figure 6.5. Note that G has an additional feature, namely, each edge is assigned a number (in this case, the number measures (roughly) the length of the corresponding street).

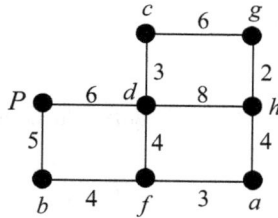

Figure 6.5

> A connected graph G is called a **weighted graph** if each edge e in G is assigned a positive number $w(e)$, called the weight of e.

Depending on the practical situation, the weight of an edge may be interpreted as a measure of physical distance, of time consumed, of cost, of capacity, or of some other quantity of interest.

Given a walk W in a weighted graph, the **weight** of W, denoted by $w(W)$, is the sum of the weights of the edges contained in W.

With this terminology, the above routing problem can now be formulated mathematically as follows:

> **The Postman Problem**
>
> Given a weighted graph G, find a closed walk W in G which passes through each edge of G at least once such that $w(W)$ is minimum.

Let us illustrate the above problem with the following examples. Consider the weighted graph H of Figure 6.6. Notice that each vertex in H is even, and so by Theorem 6.1, H is Eulerian and H possesses a closed walk which passes through each edge of H once and exactly once, i.e., an Euler circuit. Clearly, any such closed walk is a desired walk of minimum weight.

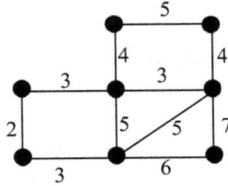

Figure 6.6

Consider, on the other hand, the weighted graph G of Figure 6.5. Notice that G contains 2 odd vertices (namely, f and h), and by Theorem 6.1, G is not Eulerian; that is, there is no closed walk in G which passes through each edge of G once and exactly once. Thus, in order to have a closed walk passing through each edge at least once, some edges in G have to be traversed more than once in the walk. But then which of the edges should a closed walk traverse more than once if the walk is of minimum weight? Clearly, this is not a simple problem.

Consider, for instance, the two closed walks in G which pass through each edge of G at least once as shown in Figure 6.7 (a) and (b).

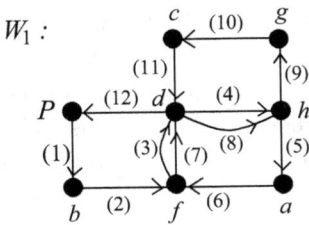

(a) edges fd and dh
are traversed twice

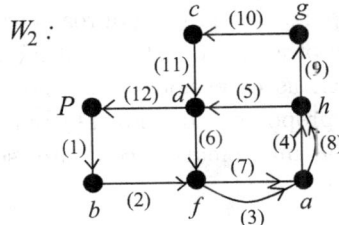

(b) edges fa and ah
are traversed twice

Figure 6.7

Note that

$$w(W_1) = 5 + 4 + 4 + 8 + 4 + 3 + 4 + 8 + 2 + 6 + 3 + 6 = 57$$

while

$$w(W_2) = 5 + 4 + 3 + 4 + 8 + 4 + 3 + 4 + 2 + 6 + 3 + 6 = 52.$$

Thus $w(W_2) < w(W_1)$. Indeed, it turns out that W_2 is a desired closed walk of minimum weight.

The Postman Problem was actually first formulated and studied around 1960 by a Chinese mathematician, Guan Meigu, who was then a Professor at Shandong University. Guan came out with a solution to the problem, and the solution was contained in an article entitled 'Graphic programming using odd or even vertices' published in the journal, *Chinese Mathematics*, in 1962. The article was translated into English and since then the problem has been called the Chinese Postman Problem.

The solution given by Guan indeed provides a procedure for finding such an optimal route. The procedure, however, is not efficient when the street network becomes large as the time required to find an optimal route could be very long. In 1973, Professor Jack Edmonds of the University of Waterloo, Canada, and Dr Ellis L. Johnson of IBM Watson Research Centre, New York, developed a new procedure for solving the Problem using some new notions on graphs. It turns out that their procedure is very much more efficient than that of Guan.

The Postman Problem has found applications in various areas such as police patrol scheduling, routings of garbage and refuse collection, street sweeping, the spraying of roads with salt-grit preventing ice formation, and many others. To reflect real life situations more closely, many variations as well as extensions and generalizations of the Postman Problem have been proposed and studied by numerous researchers from various disciplines such as mathematics, computer science, operations research, management science etc.

Exercise 6.2

(1) Determine whether the following multigraphs are Eulerian, semi-Eulerian or neither:

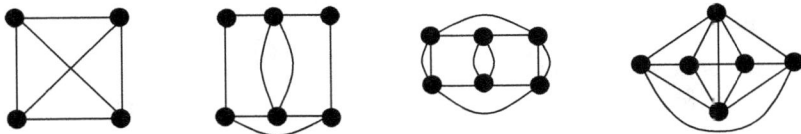

(2) Let G be the multigraph considered in Problem 1 of Exercise 6.1. Does G contain a circuit with 7 edges? Justify your answer.

(3) Let G be an Eulerian multigraph of size m. Can G contain a circuit with $m - 1$ edges? Justify your answer.

(4) Determine if each of the following statements is true:

   (i) If G is an Eulerian graph, then G contains no cut-vertices.
   (ii) If G is an Eulerian graph, then G contains no bridges.
   (iii) If G is an Eulerian graph of odd order and $\overline{G}$ is connected. then $\overline{G}$ is also an Eulerian graph.

(5) Which $K(p,q)$, $p \geq q \geq 1$, are semi-Eulerian?

(6) The following multigraph is semi-Eulerian as it contains exactly two odd vertices, namely, x and z.

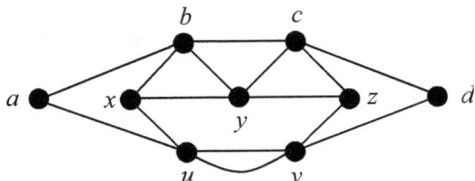

   (i) Form a new multigraph G^* by adding to G a new edge joining x and z. Is G^* Eulerian?
   (ii) Find an Euler circuit W in G^*.
   (iii) Delete xz (or zx) from W. Can you find an open Euler trail of G from the resulting sequence of edges?

(7) (+) Let G be a connected multigraph in which every vertex is even. Prove that G is Eulerian.

(8) (+) Prove Theorem 6.2.

(9) Two halls are partitioned into small rooms for an exhibition event in two different ways as shown in (a) and (b) below, where A is the entrance and B is the exit.

  (i) Is it possible for a visitor to have a route which enters at A, passes through each door once and exactly once and exits at B in partition (a)?

  (ii) Explain why such a route is not available in partition (b). Which door should be closed to ensure the existence of such a route?

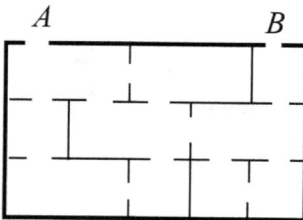
(a)

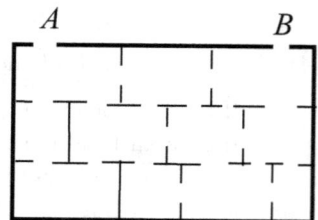
(b)

(10) (+) We have shown that the multigraph G of Figure 6.1 (a) is Eulerian. Look at its edge set $E(G)$ and observe that the edges in G can be partitioned into three edge-disjoint cycles as shown below:

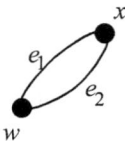
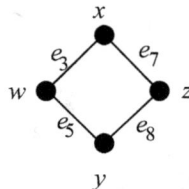
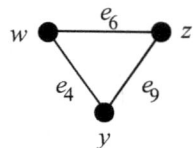

Show that, in general, a connected multigraph is Eulerian if and only if all its edges can be partitioned into some edge-disjoint cycles.

(11) Let G_1 and G_2 be two semi-Eulerian multigraphs.

    (i) Is it possible to form a semi-Eulerian multigraph by adding a new edge joining a vertex u in G_1 and a vertex v in G_2 as shown below? If the answer is 'yes', how can this be done?

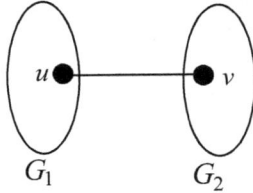

    (ii) Is it possible to form an Eulerian multigraph by adding two new edges, each of which joining a vertex in G_1 and a vertex in G_2? If the answer is 'yes', how can this be done?

(12) (+) Let G_1 and G_2 be two connected multigraphs having $2p$ and $2q$ odd vertices respectively, where $1 \leq p \leq q$. We wish to form an Eulerian multigraph from G_1 and G_2 by adding new edges, each of which joining a vertex in G_1 and a vertex in G_2. What is the least number of edges that should be added? How can this be done?

(13) (+) The following graph H is not Eulerian. What is the least number of new edges that should be added to H so that the resulting multigraph becomes Eulerian? In how many ways can this be done?

H:

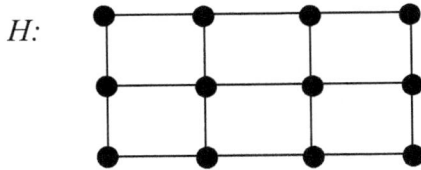

(14) Let G be a semi-Eulerian multigraph of order 8 and size 18, and with $\delta(G) = 3$ and $\Delta(G) = 6$. Assume that G contains exactly two vertices of degree 6. How many vertices of degree 3 does G have? Justify your answer and construct one such multigraph.

(15) (+) Let G be a non-trivial connected multigraph. For $A \subset V(G)$, let $e(A, V(G)\backslash A)$ denote the number of edges in G having an end in A and the other in $V(G)\backslash A$ (see Problem 29 of Exercise 2.3). Show that G is

Eulerian if and only if $e(A, V(G)\backslash A)$ is even for every proper subset A of $V(G)$.

(16) (+) Let G be a graph which contains $K(5,6)$ as a spanning subgraph.

  (i) If G is semi-Eulerian, find the minimum size of G, and construct one such extremal semi-Eulerian graph G.

  (ii) If G is Eulerian, find the minimum size of G, and construct one such extremal Eulerian graph G.

(17) (+) Let G be a multigraph which contains $K(5,7)$ as a spanning subgraph.

  (i) Assume that G is semi-Eulerian. Can G be simple? If 'yes', find the minimum size of G, and construct one such extremal semi-Eulerian graph G.

  (ii) Assume that G is Eulerian. Prove that G cannot be simple. Find the minimum size of G, and construct one such extremal Eulerian multigraph G.

(18) (+) Let G be an Eulerian graph of order 8 and size 10.

  (i) Let k be the maximum possible value of $\Delta(G)$. Determine k and construct all such G with $\Delta(G) = k$.

  (ii) Suppose that $\Delta(G) = 4$.

  (a) Determine the number of vertices of degree 4 in G.

  (b) Assume further that no two vertices of degree 4 are adjacent. Construct all such G.

6.3   Around the world and Hamiltonian graphs

In 1857, Sir William Rowan Hamilton (1805 – 1865), who was a Royal Astronomer of Ireland, invented a game called the **Icosian Game**. The game involved a regular dodecahedron (see Figure 6.8 (a)) on which the twenty vertices were labelled by the names of twenty cities in the world. The object of the game was to travel 'Around the World' by finding a 'walk' in the dodecahedron which visited each city once and exactly once, starting and terminating at the same city.

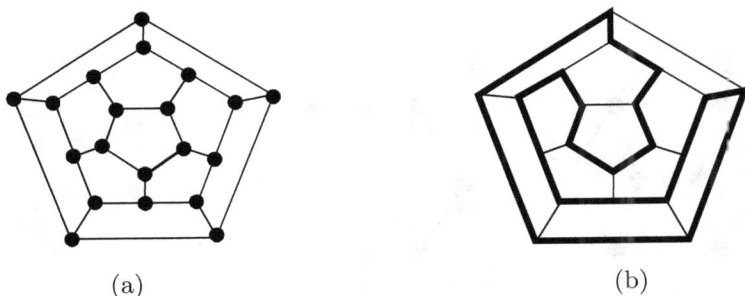

(a)   (b)

Figure 6.8

One such walk is shown in Figure 6.8(b). It is called a **spanning cycle** of the graph of Figure 6.8(a). It is a cycle because the walk is closed and, other than starting and terminating vertices, no vertex is repeated. It is spanning because all the twenty vertices of the graph are included in the walk.

> A connected graph of order $n \geq 3$ is called a **Hamiltonian graph** if it contains a spanning cycle.
>
> If G is a Hamiltonian graph, then any spanning cycle of G is called a **Hamiltonian cycle** of G.

Remark. While the Eulerian notion applies to multigraphs, we confine ourselves to (simple) graphs for the Hamiltonian notion, because if a cycle contains an edge uv, then no other parallel edge of uv can be contained in the cycle.

Question 6.3.1.

(i) Is every cycle Hamiltonian?

(ii) If a graph G of order n is Hamiltonian, what is the least possible value of $e(G)$?

(iii) Is K_n, $n \geq 3$, Hamiltonian?

(iv) Which $K(p, q)$, where $p \geq q \geq 2$, are Hamiltonian?

(v) Which of the graphs in Figure 6.9 are Hamiltonian?

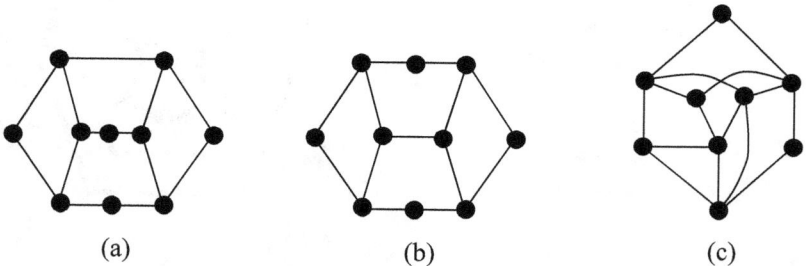

(a)   (b)   (c)

Figure 6.9

Any vertex of degree two (if such exists) has an important role to play in checking whether a given graph G is Hamiltonian or not. This is because if C is a Hamiltonian cycle of G and v is a vertex of degree two in G, then in order to visit v along C, the two edges incident with v must all be contained in C, one for entering v and one for leaving v.

This observation together with another two relevant ones are stated below.

Rules for constructing Hamiltonian cycles in G

(1) If v is a vertex of degree two in G, then the two edges incident with v are required to be contained in any Hamiltonian cycle of G.

(2) During construction, if a non-spanning cycle is formed by those required edges, then G is not Hamiltonian.

(3) If, during construction, two edges incident with a vertex are required,

then all the other edges (if there are) incident with the vertex should be deleted from further consideration.

We shall now give examples to illustrate how these rules are used to check whether or not a given graph is Hamiltonian.

Example 6.3.1. *Consider the graph G of Figure 6.9(a). For convenience, we name its vertices as shown in Figure 6.10.*

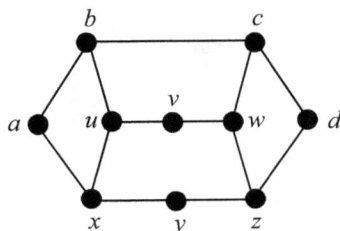

Figure 6.10

There are four vertices of degree two in G, namely, a, d, v and y. If G has a Hamiltonian cycle C, then by Rule (1), the eight edges $ab, ax, cd, dz, uv, vw, xy$ and yz are required edges in $E(C)$ as indicated in bold below:

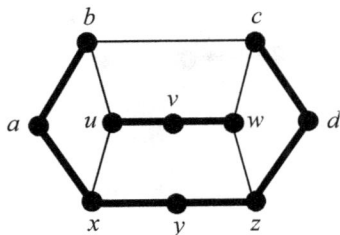

It is now obvious that the set of eight required edges can be expanded to form such a C by adding bu and cw. Thus G is a Hamiltonian graph.

Example 6.3.2. *Consider the graph G of Figure 6.9(b). Again, we name its vertices as shown in Figure 6.11.*

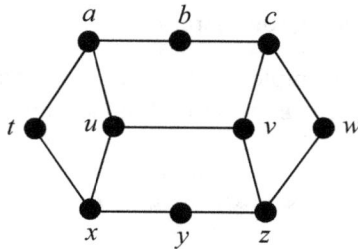

Figure 6.11

Note that G has four vertices of degree two, namely, b, t, w and y. By Rule (1), if G has a Hamiltonian cycle C, then the eight edges: $ab, bc, at, tx, cw, wz, xy$ and yz are required edges in $E(C)$ as indicated in bold below:

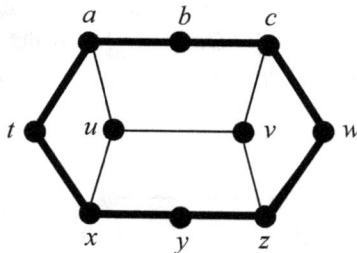

Observe that these required edges form by themselves a cycle (a C_8), which however is non-spanning. Thus, by Rule (2), G is not Hamiltonian.

Example 6.3.3. *Consider the graph G of Figure 6.9(c) and name its vertices as shown in Figure 6.12.*

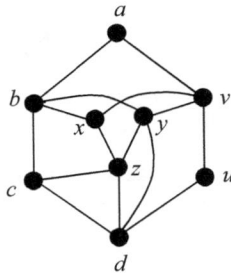

Figure 6.12

There are two vertices of degree two in G, namely, a and u. Thus, if G has a Hamiltonian cycle C, by Rule (1), the four edges: ab, va, vu and ud are required edges in $E(C)$ as indicated in bold below:

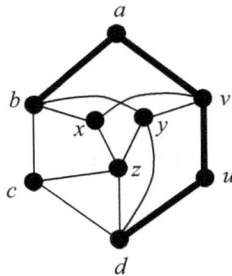

Look at the vertex v, which is of degree four.

As two of its incident edges (namely, va and vu) are required edges, by Rule (3), the other two of its incident edges (namely, vx and vy) should be deleted as shown below from further consideration.

G':

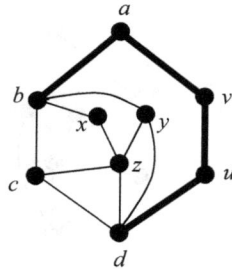

Notice that the vertex x in the resulting graph G' is now of degree two. Thus, by Rule (1), the edges xb and xz are required edges in $E(C)$ as indicated in bold below:

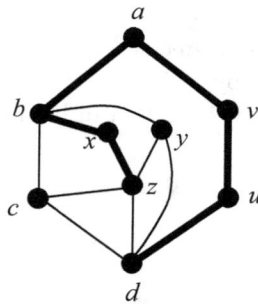

Look at the vertex b, which is of degree four in G'. As two of its incident edges, namely, ba and bx are in $E(C)$, by Rule (3), the other two incident edges, namely, bc and by should be deleted as shown below from further consideration.

G'':

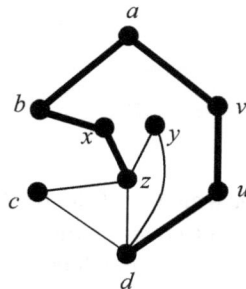

Notice that the vertex c (or y) is now of degree two in the resulting graph G''. By Rule (1), the edges: cz and cd are required edges in $E(C)$.

Finally, observe that these required edges form by themselves a non-spanning cycle (indeed, a C_8) as shown in bold below. Thus, by Rule (2), G is not Hamiltonian.

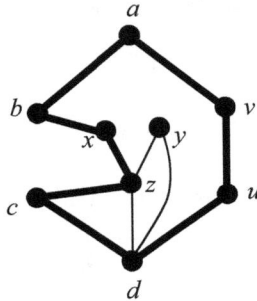

6.4   A necessary condition for a graph to be Hamiltonian

Theorem 6.1 provides us with a nice characterization of Eulerian multigraphs. In contrast with this, up till now, no good characterization of Hamiltonian graphs has been found. Indeed, the problem of characterizing Hamiltonian graphs is very hard, and is considered as one of the major unsolved problems in Graph Theory.

In this section, we shall establish a result which gives a necessary condition for a graph to be Hamiltonian. As we shall witness, this result is very useful to show that certain graphs are not Hamiltonian.

The concept of components of a disconnected graph was introduced in Section 1.4.

Let $c(H)$ denote the number of components of a graph H. (See Problem 11 of Exercise 2.3.)

Thus, for the graph H shown below, $c(H) = 5$.

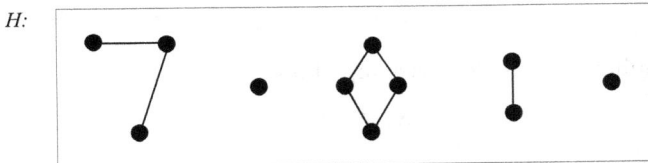

Question 6.4.1.   *Let G be the graph of Figure 6.13.*

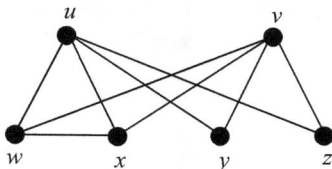

Figure 6.13

(i) Let $S = \{u, v\}$. Draw the graph $G - S$.

(ii) Find $|S|$ and $c(G - S)$.

(iii) Is $|S| < c(G - S)$?

(iv) Is G Hamiltonian?

Theorem 6.3. *Let G be a graph. If G is Hamiltonian, then for any non-empty proper subset S of $V(G)$,*

$$c(G - S) \leq |S|.$$

Proof.   Suppose that G is Hamiltonian and let S be any non-empty proper subset of $V(G)$. We shall show that

$$c(G - S) \leq |S|.$$

Since G is Hamiltonian, G contains a Hamiltonian cycle, say, C. Two observations are in order:

(1) As C is a spanning cycle of G, $C - S$ is a spanning subgraph of $G - S$, and we have (see Problem 11 of Exercise 2.3):

$$c(C - S) \geq c(G - S).$$

(2) As C is a cycle and $S \subset V(C)$, we have (see Problem 12 of Exercise 2.3):

$$c(C - S) \leq |S|.$$

Now, combining the inequalities in (1) and (2), we have:

$$c(G - S) \leq c(C - S) \leq |S|,$$

as required.   □

Remark. By Theorem 6.3, "$c(G - S) \leq |S|$ for any non-empty proper subset S of $V(G)$" is a necessary condition for G to be Hamiltonian. Thus, given a graph G, if we can find a subset S of $V(G)$ such that

$$c(G - S) > |S|,$$

then we can conclude by Theorem 6.3 that G is not Hamiltonian.

Example 6.4.1. *Consider the graph G of Figure 6.11. We have proved in Example 6.3.2 that G is not Hamiltonian. Let us now apply Theorem 6.3 to draw the same conclusion.*
  Let $S = \{a, c, x, z\}$. Then $G - S$ is shown below:

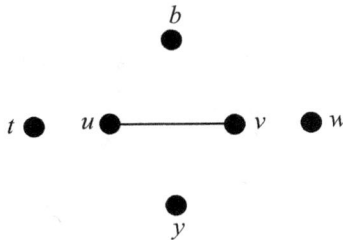

Check that $|S| = 4$ and $c(G - S) = 5$, and we have: $c(G - S) > |S|$.
  We thus conclude by Theorem 6.3 that G is not Hamiltonian.

Question 6.4.2. *Consider the graph G of Figure 6.12. Find a set S of vertices in G such that $c(G - S) > |S|$, and thus conclude that G is not Hamiltonian.*

Question 6.4.3. *A vertex w in a connected graph G is called a **cut-vertex** (see Problem 18 of Exercise 2.3) if $G - w$ is disconnected. Can G be Hamiltonian if G contains a cut-vertex? Why?*

Question 6.4.4. *The necessary condition given in Theorem 6.3 for a graph G to be Hamiltonian is, however, not sufficient. That is, the following implication is, in general, not true:*
  $c(G - S) \leq S$ for any non-empty proper subset S of $V(G)$
  $\Rightarrow G$ is Hamiltonian.
  Find a graph G such that $c(G - S) \leq |S|$ for any non-empty proper subset S of $V(G)$, but G is not Hamiltonian.

Exercise 6.4

(1) Determine whether the following graphs are Hamiltonian. Justify your answers.

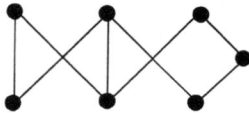
(a)

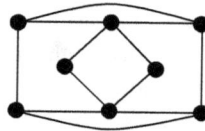
(b)

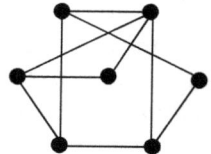
(c)

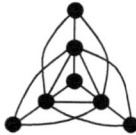
(d)

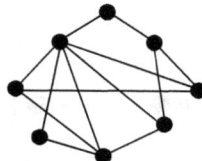
(e)

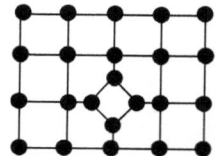
(f)

(2) Determine whether the following $m \times n$ rectangular grids are Hamiltonian.

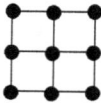
(i) 3×3

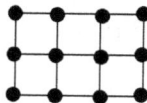
(ii) 3×4

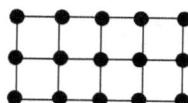
(iii) 3×5

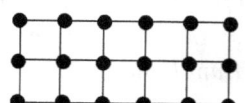
(iv) 3×6

(3) (+) Show that an $m \times n$ rectangular grid is Hamiltonian if and only if either m or n is even.

(4) Let H be a spanning subgraph of a graph G. Which of the following statements is/are true?

  (i) If H is Eulerian, then G is Eulerian.
  (ii) If H is semi-Eulerian, then G is semi-Eulerian.
  (iii) If H is Hamiltonian, then G is Hamiltonian.

(5) Prove that if G is a Hamiltonian graph, then $d(v) \geq 2$ for each vertex v in G.

(6) (i) Let H be a graph such that $d(v) = 2$ for each vertex v in H. Is H Hamiltonian?

(ii) Let H be a connected graph such that $d(v) = 2$ for each vertex v in H. Is H Hamiltonian?

(7) (+) Let G be a Hamiltonian graph with a Hamiltonian cycle C. For any non-empty proper subset A of $V(G)$, let $e_C(A, V(G) \backslash A)$ denote the number of edges in C having an end in A and the other in $V(G) \backslash A$. Show that $e_C(A, V(G) \backslash A)$ is always even.

(8) Consider the following graph G and let $S = \{x, y, z\}$.

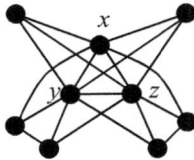

(i) Draw the graph $G - S$.
(ii) Find $|S|$ and $c(G - S)$.
(iii) Is $|S| < c(G - S)$?
(iv) Is G Hamiltonian?

(9) Let H be the graph depicted below:

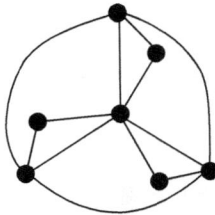

(i) Verify that $c(H - S) \le |S|$ for all non-empty proper subsets S of $V(H)$.
(ii) Is H Hamiltonian?
(iii) Is the converse of Theorem 6.3 true?

(10) Let G be a Hamiltonian graph of order n and size m such that $m = n+2$ and $n \ge 5$.

(i) Prove that $2 \le d(x) \le 4$ for each vertex x in G.

(ii) If G is also semi-Eulerian, what can be said about the structure of G?

(11) Let G be a semi-Eulerian and Hamiltonian graph with $v(G) = 12$, $e(G) = 17$, $\delta(G) = 2$ and $\Delta(G) = 4$.

(i) How many vertices of degree 3 can G have?

(ii) How many vertices of degree 2 does G have?

(iii) Construct one such graph G.

(iv) Assume that the odd vertices are adjacent in G, but no two vertices of degree 2 are adjacent in G. What can be said about the structure of the subgraph induced by the set of vertices of degree 4 in G?

(12) Let H be a semi-Eulerian and Hamiltonian graph with a Hamiltonian cycle C. Assume that $v(H) = 7$, $e(H) = 12$, $\delta(H) = 2$ and $\Delta(H) = 5$, and that H has exactly 2 vertices of degree 2.

(i) Find the number of vertices of degree 4 and the number of vertices of degree 5 in H.

(ii) Assume that the 2 vertices of degree 2 are adjacent in C. Construct all such graphs H.

(13) (+) Let G be a Hamiltonian bipartite graph of order 8.

(i) Explain why $\delta(G) \geq 2$ and $\Delta(G) \leq 4$.

(ii) Assume further that G is Eulerian and $\Delta(G) = 4$. What can be said about the structure of G?

(14) (+) Let H be the graph given below:

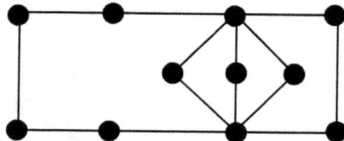

(i) Is H Hamiltonian? Why?

(ii) Let $m(H)$ denote the minimum number of new edges that are needed to add to H so that the resulting graph H^* is Hamiltonian (note that $V(H^*) = V(H)$). Find the value of $m(H)$ and justify your answer.

(iii) Construct two such Hamiltonian graphs H^* obtained by adding $m(H)$ new edges to H.

(15) (+) Let H be the graph given below:

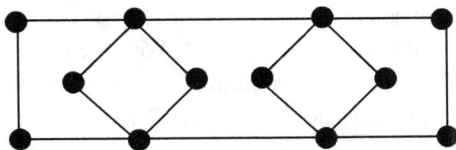

  (i) Is H Hamiltonian? Why?

 (ii) Let $m(H)$ denote the minimum number of new edges that are needed to add to H so that the resulting graph H^* is Hamiltonian (note that $V(H^*) = V(H)$). Find the value of $m(H)$ and justify your answer.

(iii) Construct two such Hamiltonian graphs H^* obtained by adding $m(H)$ new edges to H.

6.5 Two sufficient conditions for a graph to be Hamiltonian

Theorem 6.3 gives a necessary condition for a graph G to be Hamiltonian, but this condition is not sufficient. We now look for a condition which is sufficient for G to be Hamiltonian; that is, we wish to find a condition $(\#)$, say, so that if G satisfies $(\#)$, then G is Hamiltonian.

Complete graphs K_n, $n \geq 3$, are Hamiltonian. This fact, however, does not surprise us because these graphs are the 'densest' ones (i.e., containing the most number of edges for a given n), and so the task of forming a Hamiltonian cycle is easy. Must a graph G be that 'dense' in order to possess a Hamiltonian cycle? Certainly not! But then, what is the 'least density' expected of G to ensure the existence of a Hamiltonian cycle in G?

One of the most relevant quantities to measure the 'density' of a graph is the degree of a vertex. From this perspective, Gabriel Andrew Dirac (1925 – 1984) proved in 1952 the following very first significant result on Hamiltonian graphs.

Theorem 6.4. *Let G be a graph of order $n \geq 3$. If $d(v) \geq n/2$ for each vertex v in G, then G is Hamiltonian.* □

Question 6.5.1. *Let H be a 51-regular graph of order 102. Is H Hamiltonian? Why?*

Question 6.5.2. *A graph H has its degree sequence $(7, 7, 6, 6, 5, 5, 4, 4)$. Is H Hamiltonian? Why?*

Question 6.5.3. *Consider the following graph H, and let $n = v(H)$.*

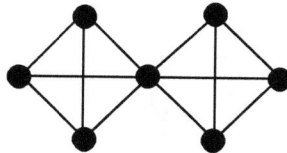

(i) Is it true that $d(v) \geq n/2$?

(ii) Is it true that $d(v) \geq (n-1)/2$?

(iii) Is H Hamiltonian?

Question 6.5.4.   *Is the converse of Theorem 6.4 true? That is, if G is a Hamiltonian graph of order n, is it true that $d(v) \geq n/2$ for each vertex v in G?*

The condition imposed on a graph G in Dirac's theorem to ensure that G contains a Hamiltonian cycle is based on the degree of each individual vertex in G. In 1960, Oystein Ore (1899 – 1968) found another sufficient condition which applies to pairs of non-adjacent vertices rather than single ones as shown below.

Theorem 6.5. *Let G be a graph of order $n \geq 3$. If*
$$d(u) + d(v) \geq n$$
for every pair of non-adjacent vertices u and v in G, then G is Hamiltonian.
□

Question 6.5.5.   *Is the converse of Theorem 6.5 true? That is, if G is a Hamiltonian graph of order $n \geq 3$, is it true that $d(u) + d(v) \geq n$ for every pair of non-adjacent vertices u, v in G?*

Question 6.5.6.   *Consider the following graph G, and let $n = v(G)$.*

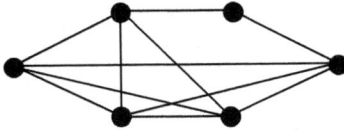

(i) *Is it true that $d(v) \geq n/2$ for each vertex v in G?*

(ii) *Is it true that $d(u) + d(v) \geq n$ for every pair of non-adjacent vertices u, v in G?*

(iii) *Can you conclude that G is Hamiltonian by Theorem 6.5?*

(iv) *Can you conclude that G is Hamiltonian by Theorem 6.4?*

Question 6.5.7.   *Consider the following graph G, and let $n = v(G)$.*

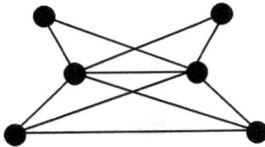

(i) *Is it true that $d(u) + d(v) \geq n$ for every pair of non-adjacent vertices u, v in G?*

(ii) *Is G Hamiltonian?*

A Related Problem – The Travelling Salesman Problem (TSP)

Mr. Tan is a travelling salesman. He has customers in 4 cities A, B, C and D, and is planning a sales trip to visit each of his customers. Mr. Tan needs to start and end the trip at his own town H. Figure 6.14 shows the cost of a one-way airline ticket between each pair of cities. If Mr. Tan chooses the route: $HABCDH$, the travelling cost is $83 + 135 + 71 + 124 + 149 = 562$. If he takes the route: $HACBDH$, the cost is $83 + 69 + 71 + 100 + 149 = 472$. What is the cheapest route?

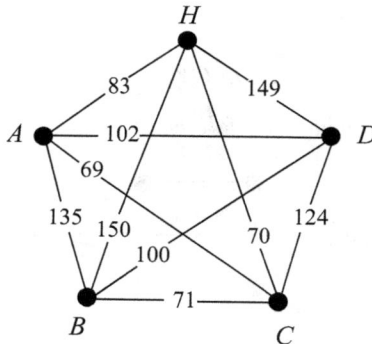

Figure 6.14

The above problem is an instance of a more general problem, known as the travelling salesman problem (TSP), which is stated below.

> Let G be a weighted complete graph. Find in G a Hamiltonian cycle with minimum weight.

The complete graph K_n contains $(n-1)!/2$ Hamiltonian cycles. For instance, if $n = 16$, then K_{16} contains $653, 837, 184, 000$ Hamiltonian cycles.

Thus, one can imagine how tough it is to solve the problem by listing. Unlike the CPP, until now, no efficient algorithm for solving the TSP is known.

The TSP has found applications in many areas such as package deliveries, fabricating circuit boards, scheduling jobs on a machine, computing wiring, dartboard design, crystallography, etc. For more relevant information on TSP, the reader may refer to the book [Excursions in Modern Mathematics] by Tannenbaum and Arnold.

Exercise 6.5

(1) The following graph H is Hamiltonian.

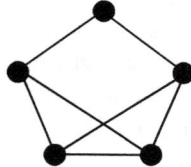

    (i) Does the Hamiltonicity of H follow from Theorem 6.4?

    (ii) Does the Hamiltonicity of H follow from Theorem 6.5?

(2) A graph G has $(8,8,8,7,7,7,6,5,5,5)$ as its degree sequence. Is G Hamiltonian? Why?

(3) (+) Let G be a graph of order $n \geq 3$. The sufficient condition given by Dirac in Theorem 6.4 states that

    (D) $d(v) \geq n/2$ for each v in $V(G)$.

The sufficient condition given by Ore in Theorem 6.5 states that

    (O) $d(u) + d(v) \geq n$ for every pair of u, v in $V(G)$ with $uv \notin E(G)$.

(1) Which of the following implications is true?

    (i) **(D)** $\Rightarrow$ **(O)**;

    (ii) **(O)** $\Rightarrow$ **(D)**.

(2) Which of the following implications is true?

    (i) Theorem 6.4 $\Rightarrow$ Theorem 6.5;

    (ii) Theorem 6.5 $\Rightarrow$ Theorem 6.4.

(4) (+) Let G be a graph of order $n \geq 3$ and size m.

    (i) Assume that there exist two non-adjacent vertices u and v in G such that $d(u) + d(v) \leq n - 1$. Show that $m \leq \binom{n-1}{2} + 1$.

    (ii) Deduce that if $m \geq \binom{n-1}{2} + 2$, then G is Hamiltonian.

(5) Construct a non-Hamiltonian graph of order $n \geq 3$ with size $\binom{n-1}{2} + 1$.

(6) (+) A path in a graph G is called a **Hamiltonian path** if it includes all the vertices in G.

    (i) Is the following graph Hamiltonian? Does it contain a Hamiltonian path?

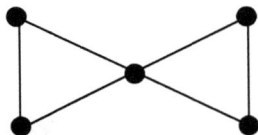

(ii) Prove that if G is a graph of order $n \geq 2$ such that $\delta(G) \geq (n-1)/2$, then G contains a Hamiltonian path.

(7) (+) For each odd integer $n \geq 3$, construct a non-Hamiltonian graph G of order n such that $\delta(G) = (n-1)/2$.

(8) (+) Let $G + H$ denote the **join** of two graphs G and H (see Problem 27 of Exercise 4.3). For a positive integer r, denote by rK_2 the union of r independent edges as shown below:

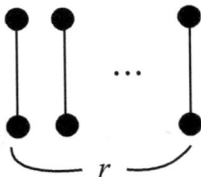

(i) Determine whether the join $(3K_2) + N_7$ is Hamiltonian. Justify your answer.
(ii) Determine whether the join $(4K_2) + N_7$ is Hamiltonian. Justify your answer.

Chapter 7

Digraphs and Tournaments

7.1 Digraphs

The multigraph G of Figure 7.1 (a) models the street network of the business section of a town.

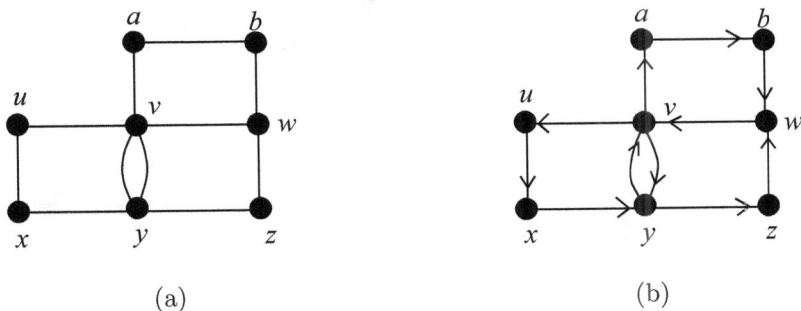

(a)   (b)

Figure 7.1

As the traffic in this section has increased significantly, the town council has decided to convert the current two-way traffic system in this section to a one-way system, and one possible solution is depicted in Figure 7.1(b), where an arrow of an edge indicates the proposed direction of the corresponding street.

The diagram shown in Figure 7.1 (b) is called a **directed graph** or, in short, a **digraph**. Like a multigraph, a digraph has a set of vertices; but a digraph differs from a multigraph in that any edge which links two vertices has a 'direction'.

A **digraph**, D, consists of a non-empty finite set $V(D)$ together with a set $E(D)$ of **ordered** pairs of distinct elements of $V(D)$.

The set $V(D)$ is called the **vertex set** of D and the set $E(D)$ of 'directed edges' is called the **arc set** of D. While the elements in $V(D)$ are the **vertices** in D, those in $E(D)$ are called the **arcs** in D.

The **order** of D, denoted by $v(D)$, is the number of vertices in D; that is, $v(D) = |V(D)|$. The **size** of D, denoted by $e(D)$, is the number of arcs in D; that is, $e(D) = |E(D)|$.

Example 7.1.1.   *Let D be the digraph shown in Figure 7.1(b). Then*

$$V(D) = \{a, b, u, v, w, x, y, z\}$$

and

$E(D) =$

$\{(a, b), (b, w), (u, x), (v, a), (v, u), (v, y), (w, v), (x, y), (y, v), (y, z), (z, w)\}.$

Also, the order of D is 8 (that is, $v(D) = 8$) and the size of D is 11 (that is, $e(D) = 11$).

In Example 7.1.1, the **ordered** pair of vertices, (a, b), represents the arc:

In this case, (1) we may write ab or $a \to b$ for (a, b);
(2) we may say that

  (i) a is adjacent to b,

 (ii) b is adjacent from a,

(iii) a dominates b,

(iv) b is dominated by a,

 (v) the arc ab is incident from a,

(vi) the arc ab is incident to b.

Remark. As **ordered** pairs, $(p, q) \neq (q, p)$ in general. In the digraph D of Figure 7.1(b), we have the arc (a, b), but not (b, a); we have both (v, y) and (y, v), but they represent different arcs.

Question 7.1.1. *Let D be the digraph depicted in Figure 7.2.*

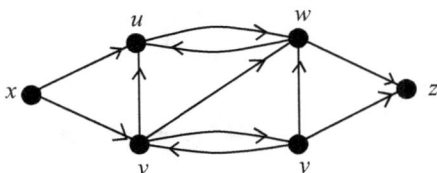

Figure 7.2

(i) *Find $V(D)$ and $E(D)$.*

(ii) *Find $v(D)$ and $e(D)$.*

(iii) *Which vertices are dominated by v?*

(iv) *Which vertices dominate w?*

(v) *Which vertices are adjacent from x?*

(vi) *Which vertices are adjacent to z?*

(vii) *Which arcs are incident to u?*

(viii) *Which arcs are incident from y?*

Question 7.1.2. *Let D be the digraph defined as follows:*

$$V(D) = \{a, b, c, d, e, f\}$$

and

$$E(D) = \{ab, ad, af, bc, bd, be, bf, ca, cb, ce, db, de, ea, ec, ef, fa, fc\}.$$

(i) *Draw the diagram of D.*

(ii) *Find $v(D)$ and $e(D)$.*

(iii) *Which vertices are dominated by a?*

(iv) *Which vertices dominate a?*

(v) *Which vertices are adjacent to f?*

(vi) Which vertices are adjacent from f?

(vii) Which arcs are incident to e?

(viii) Which arcs are incident from e?

Remark. Consider the diagram depicted in Figure 7.3.

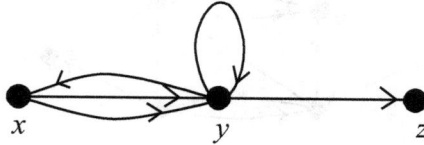

Figure 7.3

There are two arcs incident from x to y. We call them parallel arcs. Also, there is an arc incident from y to y. We call it a loop. Throughout this chapter, we shall assume the following:

All digraphs contain neither parallel arcs nor loops.

Exercise 7.1

(1) Let D be the digraph shown below:

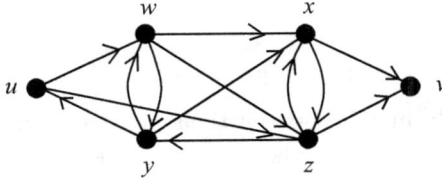

Find

    (i) $V(D)$ and $E(D)$;

    (ii) $v(D)$ and $e(D)$;

   (iii) all vertices adjacent from w;

   (iv) all vertices adjacent to y;

    (v) all vertices dominated by x;

   (vi) all vertices that dominate z;

  (vii) all arcs incident from u;

 (viii) all arcs incident to z.

(2) Let D be the digraph defined as follows:

$$V(D) = \{a, b, u, v, w, x, y, z\}$$

and

$$E(D) = \{aw, ay, bx, ux, vz, wu, wx, wy, xz, ya, yv, yx, yz, zb, zy\}.$$

    (i) Draw the diagram of D.

    (ii) Find $v(D)$ and $e(D)$.

   (iii) Which vertices are adjacent from a?

   (iv) Which vertices are adjacent to y?

    (v) Which vertices are dominated by w?

   (vi) Which vertices dominate x?

  (vii) Which arcs are incident to z?

 (viii) Which arcs are incident from y?

(3) Let D be the digraph defined as follows:

$$V(D) = \{1, 2, 3, 4, 5, 6\}$$

and $(i, j) \in E(D)$, where i, j are in $V(D)$, if and only if $i > j$.

    (i) Draw the diagram of D.

  (ii) Find $e(D)$.

 (iii) Which vertices are adjacent to '2'?

 (iv) Which vertices are adjacent from '2'?

(4) Two table tennis teams A and B, each consisting of 3 players as shown below:

$$A = \{x, y, z\} \text{ and } B = \{u, v, w\},$$

had a friendly match between their players in singles. Each player in a team must play each player in the other exactly once with no ties allowed. At the end of the match, it was reported that

  (i) x won all the matches;

 (ii) y was defeated only by w;

(iii) team A defeated team B by just one match.

(1) Construct a digraph D to model the situation where $V(D)$ is the set of all players, and $a \to b$ in D if player a defeated player b.

(2) Find $e(D)$.

(3) Did z win any game?

(4) Which player in team B won the largest number of matches?

7.2   Basic concepts

In this section, we shall introduce a number of basic concepts in digraphs which are parallel to their counterparts in multigraphs.

The in-degree and out-degree of a vertex

The degree of a vertex in a multigraph is defined as the number of edges incident with the vertex. Due to the presence of **directions** of the arcs, we have two different types of degrees for vertices in a digraph.

Let D be a digraph, and v a vertex in D.
The **in-degree** of v, denoted by $id(v)$, is the number of arcs adjacent to v in D.
The **out-degree** of v, denoted by $od(v)$, is the number of arcs adjacent from v in D.
A vertex in D is called a **source** if its in-degree is zero.
A vertex in D is called a **sink** if its out-degree is zero.

Question 7.2.1.   *Let D be the digraph of Figure 7.1(b). Complete the following table:*

Vertex	In-degree	Out-degree
a		
b		
u		
v		
w		
x		
y		
z		
Total Sum		

Does D contain any source or sink?

Can you see any relationship between $e(D)$ and the two total sums you found in the above table?

Question 7.2.2.   *Let D be the digraph of Figure 7.2. Complete the following table:*

Vertex	In-degree	Out-degree
u		
v		
w		
x		
y		
z		
Total Sum		

Does D contain any source or sink?

Can you see any relationship between $e(D)$ and the two total sums you found in the above table?

Question 7.2.3.   *Let D be the digraph defined in Question 7.1.2. Complete the following table:*

Vertex	In-degree	Out-degree
a		
b		
c		
d		
e		
f		
Total Sum		

Does D contain any source or sink?

Can you see any relationship between $e(D)$ and the two total sums you found in the table above?

Indeed, as an arc in a digraph contributes exactly one towards the total sum of the in-degrees and exactly one towards the total sum of the out-degrees, we have immediately the following result for digraphs, which is analogous to the Euler's handshaking lemma in multigraphs.

Result (1). Let D be a digraph. Then

$$\sum_{v \in V(D)} id(v) = e(D) = \sum_{v \in V(D)} od(v).$$

Isomorphic digraphs

Just like the situation for graphs, given two digraphs, we may wish to know whether they are the 'same'. This leads to the following concept.

Let D_1 and D_2 be two digraphs. We say that D_1 is **isomorphic** to D_2, and we write $D_1 \cong D_2$, if there exists a one-one and onto mapping $f : V(D_1) \to V(D_2)$ such that $(u, v) \in E(D_1)$ if and only if $(f(u), f(v)) \in E(D_2)$, where u, v are vertices in D (that is, the dominance relation is preserved under f). Such a one-one and onto mapping f is called an **isomorphism** from D_1 to D_2.

Example 7.2.1. *Let D_1 and D_2 be the two digraphs shown in Figure 7.4.*

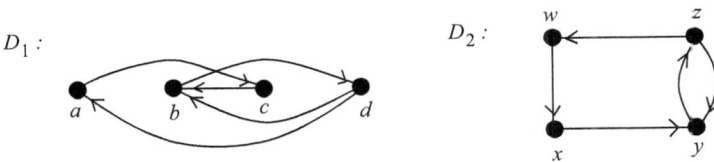

Figure 7.4

We claim that $D_1 \cong D_2$. Indeed, define the mapping $f : V(D_1) \to V(D_2)$ as follows: $f(a) = w, f(b) = y, f(c) = x$ and $f(d) = z$.

Clearly, f is both one-one and onto. Moreover, it can be checked that f preserves the dominance relation (for instance, $a \to c$ and $f(a) \to f(c)$, $d \to b$ and $f(d) \to f(b)$, etc.). Thus, we conclude that $D_1 \cong D_2$.

Question 7.2.4. *There are three digraphs of the same order and same size shown in Figure 7.5. Which two are isomorphic? Justify your answer.*

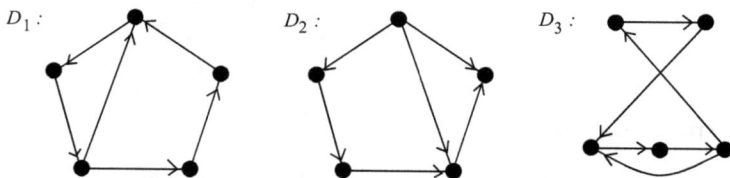

Figure 7.5

Connectedness

Let D be a digraph and u, v be vertices in D.

(1) A $u - v$ **walk** is a finite sequence of vertices: $v_0 v_1 \cdots v_k$, where $v_0 = u$, $v = v_k$, and for each $i = 0, 1, \cdots, k - 1$, $v_i \to v_{i+1}$ in D. The **length** of the walk $v_0 v_1 \cdots v_k$ is k, which is the number of occurrences of arcs in the walk.

(2) A $u - v$ walk is **closed** (resp., **open**) if $u = v$ (resp., $u \neq v$).

(3) A $u - v$ **path** is a $u - v$ walk in which no vertex is repeated.

(4) A **cycle** is a closed walk: $v_1 v_2 \cdots v_k v_1$, where all $v_1, v_2, \cdots, v_k$ are distinct. Such a cycle is also called a k-**cycle**.

(5) The vertex v is **reachable from** the vertex u if there is a $u - v$ path in D.

(6) If v is reachable from u, the **distance from** u **to** v, $d(u, v)$, is the smallest length of all $u - v$ paths in D. We define $d(u, v) = \infty$ if v is not reachable from u.

(7) The **underlying graph** (or **underlying multigraph**) of D, denoted by $G(D)$, is the graph (or multigraph) obtained from D by disregarding the direction of each arc in D.

(8) The digraph D is **connected** if $G(D)$ is connected.

(9) The digraph D is **strongly connected**, or simply **strong**, if every two vertices in D are mutually reachable.

Example 7.2.2.   *We show again the digraph D of Figure 7.1(b) below for convenience.*

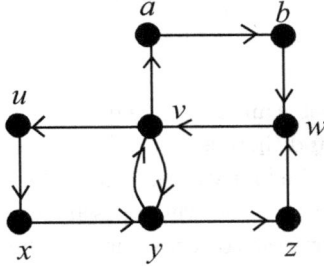

(i) *Disregarding the directions of arcs in D, we obtain its underlying multi-graph $G(D)$, which is the multigraph of Figure 7.1(a).*
    As $G(D)$ is connected, the digraph D is thus connected.

(ii) *There are only two $a - y$ paths in D, namely, $abwvuxy$ and $abwvy$. The former one is of length 6 while the latter of length 4. By definition, $d(a, y) = 4$.*

(iii) *In D, we can find a 2-cycle, namely, vyv but no 3-cycle. A number of 4-cycles are available; one of them is $abwva$. The largest cycles in D are 6-cycles; one of them is $xyzwvux$.*

(iv) *It can be checked that every two vertices in D are mutually reachable. Thus D is a strong digraph. Note that $d(v, y) = d(y, v) = 1; d(v, a) = 1$ but $d(a, v) = 3$; while $d(w, u) = 2, d(u, w) = 4$.*

Remark. We see from (iv) that the distance function d in digraphs is not symmetric.

Question 7.2.5.   *Consider the digraph D of Figure 7.2.*

(i) *Find an $x - z$ walk of length 6 in D.*

(ii) *Find an $x - z$ path of length 4 and an $x - z$ path of length 3 in D. Does there exist an $x - z$ path of length 2 in D? What is the value of $d(x, z)$?*

(iii) *Is x reachable from z? What is the value of $d(z, x)$?*

(iv) *Find two 2-cycles in D.*

(v) *Is there any k-cycle in D, where $k \geq 3$?*

(vi) Is D a strong digraph?

(vii) Draw G(D).

*(viii) In general, can a digraph be strong if it contains a source or a sink?
Why?*

The World Wide Web can be modeled as a digraph. A webpage is represented by a vertex and there is an arc from a vertex to another vertex if there is a hyperlink from the first vertex to the other vertex. There are loops as well by this definition as some webpages have links to various sections within themselves. Of interest to researchers is the diameter of the World Wide Web as this would indicate the maximum number of links needed from any given webpage to another. (Just as for graphs, we define the diameter of a digraph D as the greatest distance between any two vertices in D.) Also of interest are vertices of high degree. Vertices of high out-degree have been called 'hubs'. These are usually the search engines and pages of bookmarks. Vertices of high in-degree have been called 'authorities', i.e., many webpages have links to them thus showing the importance of their content. In fact, some search engines work on this property and rank a webpage according to its in-degree, the in-degrees of the webpages adjacent to it, the in-degrees of the webpages adjacent to them, and so on.

Exercise 7.2

(1) Let D be the digraph considered in Problem 1 of Exercise 7.1.

   (i) Find the in-degree and out-degree of each vertex in D.
   (ii) Verify that the equalities in Result (1) hold.
   (iii) Is there any source in D?
   (iv) Is there any sink in D?
   (v) For $k = 2, 3, 4, 5$, find a k-cycle in D.
   (vi) Is there any 6-cycle in D? Why?
   (vii) Which vertices are reachable from u?
   (viii) Which vertices are reachable from v?
   (ix) Is D connected?
   (x) Is D strongly connected?
   (xi) Find $d(u, x), d(u, z), d(u, v), d(x, w)$ and $d(v, x)$.

(2) Let D be the digraph considered in Problem 2 of Exercise 7.1.

   (i) Find the in-degree and out-degree of each vertex in D.
   (ii) Verify that the equalities in Result (1) hold.
   (iii) Is there any source in D?
   (iv) Is there any sink in D?
   (v) Find a 6-cycle in D.
   (vi) Is there any 7-cycle in D?
   (vii) Is there any 8-cycle in D?
   (viii) Which vertices are reachable from u?
   (ix) Is D connected?
   (x) Is D strongly connected?
   (xi) Find $d(a, u), d(u, a), d(a, b)$ and $d(b, a)$.
   (xii) Find two vertices in D such that the distance from one of them to the other is 5.
   (xiii) Find two vertices in D such that the distance from one of them to the other is 6.

(3) Let D be the digraph considered in Problem 3 of Exercise 7.1.

   (i) Find $e(D)$.
   (ii) Find the in-degree and out-degree of each vertex in D.
   (iii) Verify that the equalities in Result (1) hold.
   (iv) Is there any source in D?
   (v) Is there any sink in D?
   (vi) Are there any two vertices in D which have the same out-degree?

(vii) Is there any cycle in D?

(viii) Is there any path of length 5 in D?

(ix) What is $G(D)$?

(x) Is D strong?

(4) Are the following two digraphs isomorphic? If 'yes', find an isomorphism from one of them to the other.

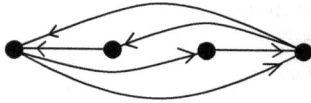

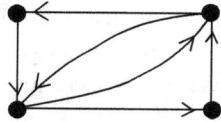

(5) Are the following two digraphs isomorphic? Justify your answer.

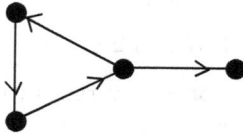

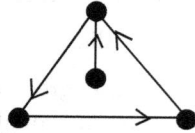

(6) Among the following three digraphs, which two are isomorphic? Justify your answer.

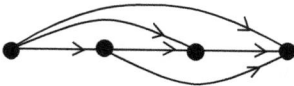

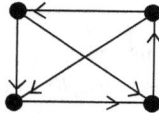

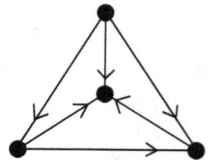

(7) (+) Let D be a digraph. Prove that every $u - v$ walk in D contains a $u - v$ path, where u and v are two vertices in D.

(8) (+) Let D be a digraph and $k = \min\{od(v)|v \in V(D)\}$. Show that D contains a path of length at least k. Is the result also true if $k = \min\{id(v)|v \in V(D)\}$?

(9) (+) Let D be a digraph and $k = \min\{od(v)|v \in V(D)\}$. Show that D contains an r-cycle, where $r \geq k + 1$. Is the result also true if $k = \min\{id(v)|v \in V(D)\}$?

(10) Let D be a digraph with $v(D) \geq 2$. Prove that if D is strong, then $id(v) \geq 1$ and $od(v) \geq 1$ for each vertex v in D (that is D contains neither sink nor source). Is the converse true?

(11) (+) Let D be a digraph whose underlying graph $G(D)$ is the following 2×3 grid:

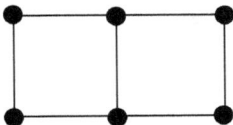

Assume that D contains neither sink nor source. Show that D contains a 4-cycle.

(12) (+) Let D be a digraph whose underlying graph $G(D)$ is $K(4,4)$. Prove that if D contains an 8-cycle, then D contains a 4-cycle.

(13) (+) Let D be a digraph of order $n \geq 2$. Assume that $od(x) + id(y) \geq n - 1$ for any two vertices x, y in D such that x is not adjacent to y. Prove that D is strong.

(14) (+) Let D be a digraph that contains no cycles. Prove that D contains a sink and a source. Is the converse true?

(15) (+) Let D be a digraph. Prove that D contains no cycles if and only if every walk in D is a path.

(16) (+) Construct a digraph D of order 7 such that $id(v) = od(v) = 2$ for each vertex v in D, but D contains no k-cycles, where $k = 2, 4, 6$.

(17) Let D be a digraph and A, a set of vertices in D. Denote by $R(A)$ the set of vertices in D which are reachable from some vertex in A Clearly, $A \subseteq R(A)$, as every vertex is reachable from itself.

(a) Consider the following digraph D:

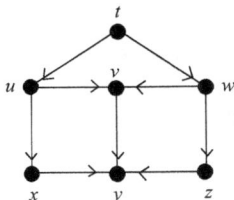

(i) Let $S = \{x, z\}$, $T = \{u, w\}$ and $U = \{x, y, z\}$. Find $R(S)$, $R(T)$ and $R(U)$.

(ii) Is $R(U) = U$?

(iii) Is D strong?

(b) Consider the following digraph D:

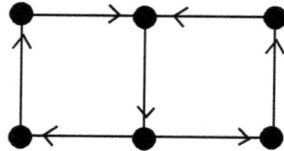

(i) Can you find a non-empty set W of vertices in D such that $W \neq V(D)$ and $R(W) = W$?

(ii) Is D strong?

(c) (+) Prove that a digraph D is strong if and only if for every non-empty and proper subset W of $V(D)$, $W \subset R(W)$.

(18)   An **orientation** of a graph G is a digraph D obtained from G by assigning each edge in G an arbitrary direction. (Clearly, the underlying graph of D is G.) For instance, a graph G and an orientation of G are shown below.

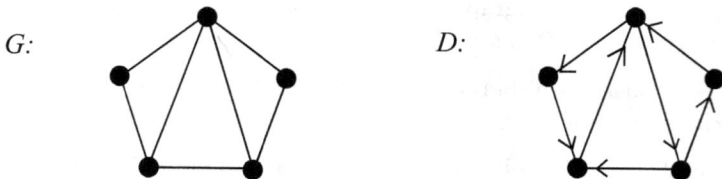

(i) Let G be a graph. Does there exist an orientation D of G such that $d(x, y) \leq 1$ for all x, y in $V(D)$?

(ii) For $n = 3, 5$ and 6, find an orientation D_n of K_n such that $d(x, y) \leq 2$ for all x, y in $V(D_n)$.

(iii) Find an orientation D_4 of K_4 such that $d(x, y) \leq 3$ for all x, y in $V(D_4)$.

(iv) (+) Does there exist an orientation D_4 of K_4 such that $d(x, y) \leq 2$ for all x, y in $V(D)$?

(v) (+) Suppose that, for some $n \geq 5$, K_n has an orientation D_n such that $d(x, y) \leq 2$ for all x, y in $V(D_n)$. Construct an orientation D_{n+2} of K_{n+2} based on D_n such that $d(x, y) \leq 2$ for all x, y in $V(D_{n+2})$.

(19) Consider the orientation D of the graph G shown at the beginning of Problem 18.

(i) Verify that $d(x, y) \leq 4$ in D for all x, y in $V(D)$.
(ii) Is it true that $d(x, y) \leq 3$ in D for all x, y in $V(D)$?
(iii) Find an orientation D' of G such that $d(x, y) \leq 3$ in D' for all x, y in $V(D')$.
(iv) Does there exist an orientation D^* of G such that $d(x, y) \leq 2$ in D^* for all x, y in $V(D^*)$? Justify your answer.

(20) Let $G = K(p, p)$, where $p \geq 2$.

(i) Does there exist an orientation D of G such that $d(x, y) \leq 2$ in D for all x, y in $V(D)$? Justify your answer.
(ii) Find an orientation D of G such that $d(x, y) \leq 3$ in D for all x, y in $V(D)$.

(21) Let D be a digraph with n vertices labeled $v_1, v_2, \cdots, v_n$. The **adjacency matrix** of D, denoted by $A(D)$, is the $n \times n$ matrix in which $a_{i,j}$, the entry in row i and column j, is 1 if there is an arc from vertex v_i to vertex v_j, and 0 otherwise. We may sometimes write $A(D) = (a_{i,j})$.
(i) Draw the digraph D which has the following adjacency matrix $A(D)$:

$$A(D) = \begin{pmatrix} 0 & 1 & 0 & 1 \\ 0 & 0 & 1 & 1 \\ 0 & 1 & 0 & 0 \\ 0 & 1 & 1 & 0 \end{pmatrix}$$

(ii) Find the adjacency matrix $A(D)$ of the following digraph D.

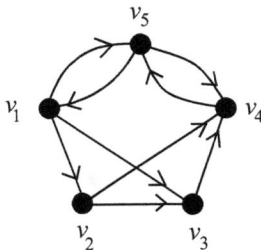

(22) Let D be a digraph of order n, where $n \geq 2$.

(i) Show that if D contains no cycles, then $od(v) = 0$ for some vertex v in D.

(ii) (+) Show that D contains no cycles if and only if the vertices in D can be named as $v_1, v_2, \cdots, v_n$ such that $A(D)$ is upper triangular. (Note that a square matrix $(a_{i,j})$ is a called an **upper triangular matrix** if $a_{i,j} = 0$ for all i, j with $i > j$.)

7.3   Tournaments

In this section, we shall introduce and study a very special type of digraphs, called tournaments.

A **tournament** is a digraph in which every two vertices are joined by exactly one arc.

Equivalently, a **tournament** is a digraph obtained by assigning a direction to each edge of a complete graph.

A tournament of order 5 is shown in Figure 7.6.

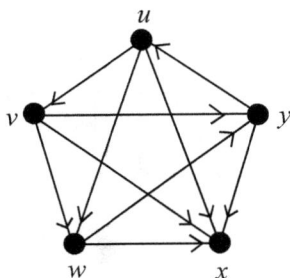

Figure 7.6

Such a mathematical model is called a tournament since it can be used to show the possible outcomes of a **round-robin tournament**. In a round-robin tournament, there is a set of players (or teams) where any two players (or teams) engage in a game that cannot end in a tie, and every player (or team) must play each other once and exactly once.

The tournament of Figure 7.6 shows the outcomes of a 5-player tennis (singles) round-robin tournament. Here, for any two players, say a and b, we denote by $a \to b$ if a defeats b. The five players in the tournament are: u, v, w, x, and y, and the results are: u defeats v, w and x, but is defeated by y, etc.

In a round-robin tournament, the **score** obtained by a player is the number of players he/she defeats, that is, the out-degree of the vertex representing the player in the tournament. Thus, in the tournament of Figure 7.6, the scores of the players u, v, w, x and y are, respectively, $3, 3, 2, 0, 2$.

Question 7.3.1.   *Five singles table tennis players a, b, c, d, e were involved in a round-robin tournament. It was reported that a defeated all other players; d was defeated only by a; and e defeated only two players. Were there any two players having the same score?*

Throughout this chapter, for a positive integer n, we shall denote by T_n a tournament of order n.

Example 7.3.1.   *For $n = 1, 2, 3, 4$, all non-isomorphic tournaments T_n are shown below.*

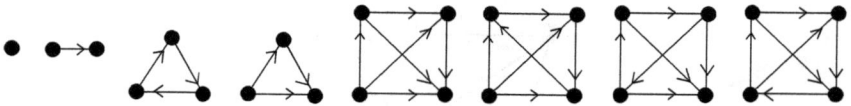

Question 7.3.2.

(i) Construct a tournament of order 5 in which every vertex has the same score.

(ii) Construct a tournament of order 5 in which no two vertices have the same score.

Some useful quantitative relations in tournaments are stated below.

Result (2).

(i) $e(T_n) = \binom{n}{2}$;

(ii) $od(v) + id(v) = n - 1$ for each vertex v in T_n; and

(iii) $\displaystyle\sum_{v \in V(T_n)} od(v) = \binom{n}{2} = \sum_{v \in V(T_n)} id(v)$.

Transitive tournaments

A special family of tournaments is introduced below.

A tournament T is said to be **transitive** if it satisfies the following
condition:

(Δ)   for any three vertices u, v, w in T, if $u \to v$ and $v \to w$, then
$u \to w$.

Question 7.3.3.   *Consider the tournaments shown in Example 7.3.1.*
Those of order 1 and 2 are, by definition, transitive. Besides these, there
are exactly two tournaments shown therein which are transitive. Which two
are transitive?

Question 7.3.4.

(i) Is the tournament you constructed in Question 7.3.2(i) transitive?

(ii) Is the tournament you constructed in Question 7.3.2(ii) transitive?

(iii) Is the tournament that models the 5-player table tennis round-robin
tournament in Question 7.3.1 transitive?

(iv) Are the two tournaments considered in (ii) and (iii) isomorphic?

Let T be a transitive tournament, and assume that we have in T:

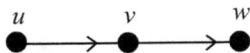

Then, by definition, we must have:

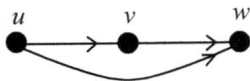

Thus, there is no way we can form a 3-cycle in T.

Indeed, we have the following result which says that transitive tournaments are precisely those tournaments which contain no cycles.

Theorem 7.1.   *Let T be a tournament. Then T is transitive if and only if T contains no cycles.*

Proof. [Necessity] Suppose T is transitive. Assume on the contrary that T contains a cycle, say,

$$x_1 \to x_2 \to \cdots \to x_k \to x_1,$$

where $k \geq 3$. Then, by applying the condition (Δ) successively, we have: $x_1 \to x_k$ and $x_k \to x_1$, a contradiction.

[Sufficiency] Suppose T contains no cycles. Assume on the contrary that T is not transitive. Then, by definition, T contains three vertices x, y, z such that $x \to y$ and $y \to z$ but $z \to x$. But then $xyzx$ is a cycle in T, a contradiction. $\square$

Exercise 7.3

(1) Let T be a tournament, and u, v be two mutually reachable vertices in T. Prove that

  (i) $d(u, v) \neq 1$ if and only if $d(v, u) = 1$;

  (ii) $d(u, v) \neq d(v, u)$.

(2) (+) Let T be a tournament, and u, v be two vertices in T with $d(u, v) = k \geq 2$. Prove that

  (i) $od(v) \geq k - 1$;

  (ii) v is contained in an r-cycle for each $r = 3, 4, \cdots, k + 1$;

  (iii) u and v are contained in a common $(k + 1)$-cycle.

(3) Show that, up to isomorphism,

  (i) there is only one strong tournament of order 3;

  (ii) there is only one strong tournament of order 4.

(4) Let T_n be a strong tournament of order n such that, for each arc in T_n, the reversing of the direction of this arc results in also a strong tournament. Show that $n \geq 5$ and construct one such T_n.

(5) (i) Suppose that six teams play in a round-robin tournament. Is it possible that all six teams have the same score at the end?

  (ii) Suppose that seven teams play in a round-robin tournament. Is it possible that all seven teams have the same score at the end?

  (iii) Suppose that n teams, $n \geq 3$, play in a round-robin tournament. Is it possible that all n teams have the same score at the end?

(6) Let T be a tournament of order $n \geq 3$. Assume that $od(v) = k$ for all vertices v in T.

  (i) Find a relation between k and n.

  (ii) Deduce that n must be odd.

  (iii) Construct one such tournament of order 7.

  Remark. Such a tournament is called a **regular tournament**.

(7) Consider the tournament T of Figure 7.6. Find

  (i) the out-degree and in-degree of each vertex in T;

  (ii) the sum $od(u) + od(v) + od(w) + od(x) + od(y)$;

  (iii) the sum $id(u) + id(v) + id(w) + id(x) + id((y)$;

  (iv) the sum $(od(u))^2 + (od(v))^2 + (od(w))^2 + (od(x))^2 + (od(y))^2$;

  (v) the sum $(id(u))^2 + (id(v))^2 + (id(w))^2 + (id(x))^2 + (id(y))^2$.

Are the sums in (ii) and (iii) the same?

Are the sums in (iv) and (v) the same?

(8) Prove the results in Result (2).

(9) (+) Let T be a tournament. Show that

$$\sum_{v \in V(T)} (od(v))^2 = \sum_{v \in V(T)} (id(v))^2.$$

(Putnam Exam (1965))

(10) (+) Let T be a tournament with $V(T) = \{v_1, v_2, \cdots, v_n\}$. Show that, for any $k = 1, 2, \cdots, n$,

$$\sum_{i=1}^{k} od(v_i) \geq \binom{k}{2}.$$

(11) Is there a tournament in which the out-degrees of the vertices are:

   (i) $5, 4, 3, 2, 1, 0$?

   (ii) $5, 5, 3, 1, 1, 0$?

   (iii) $5, 4, 4, 1, 1, 0$?

   (iv) $4, 3, 3, 2, 2, 1$?

   For each case, construct one such tournament if your answer is 'yes'; give a reason if your answer is 'no'.

(12) (i) Construct a tournament T_5 in which the out-degrees of the vertices are: 2, 2, 2, 2, 2.

   (ii) Construct a tournament T_6 in which the out-degrees of the vertices are: 3, 3, 3, 2, 2, 2.

   (iii) Show a way of combining the above T_5 and T_6 to obtain a tournament of order 11 in which the out-degrees of the vertices are: 8, 8, 8, 8, 8, 3, 3, 3, 2, 2, 2.

(13) (+) Let T be a tournament of order $n \geq 3$. Prove that T contains a 3-cycle if and only if T contains two vertices of the same out-degree.

(14) (+) Prove that a tournament T is transitive if and only if $od(u) \neq od(v)$ for any two vertices u, v in T.

(15) (+) Prove that a tournament T of order n is transitive if and only if the out-degrees of its vertices are, respectively, $n - 1, n - 2, \cdots, 1, 0$.

(16) (+) A tournament T is said to be **reducible** if $V(T)$ can be partitioned into two non-empty subsets, U and W, such that $u \to w$ for all $u \in U$ and $w \in W$. A reducible tournament of order 5 is shown in the following:

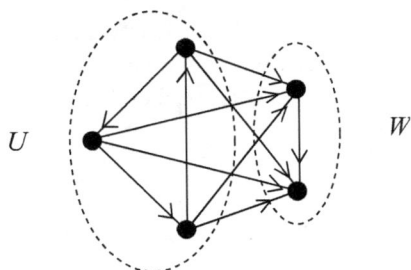

(i) Is the following tournament reducible?

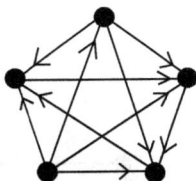

(ii) Prove that every transitive tournament is reducible.

7.4 Two properties of tournaments

Tournaments are very special digraphs, and so they are expected to have some properties that other digraphs do not have. In this section, we shall present two of them.

Consider the tournament of Figure 7.6. We observe that the five players in the round-robin tournament can be arranged in a row so that one defeats the player on his/her immediate right. For instance,

$$u \to v \to w \to y \to x, \quad y \to u \to v \to w \to x, \quad \text{etc.}$$

Such an arrangement involves all 'vertices' in a single 'path'. We call it a spanning path or Hamiltonian path of the tournament.

A path in a digraph D is called a **Hamiltonian path** if the path contains all vertices in D.

Question 7.4.1.

(i) Does the following digraph contain a Hamiltonian path?

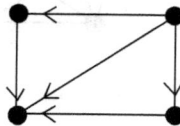

(ii) Check if each of the tournaments shown in Example 7.3.1 and Question 7.3.2 contains a Hamiltonian path.

Question 7.4.2. *For $n = 3, 4, 5, 6$, construct a transitive tournament of order n. Can you find a Hamiltonian path in each tournament? If 'yes', how many are there in each tournament?*

We are sure that you can always find such a path in answering the above two questions, except Question 7.4.1(i). Indeed, we have the following result due to Rédei [R]:

Theorem 7.2. *Every tournament contains a Hamiltonian path.*

Proof. Let T be a tournament of order n, and let

$$P: \quad x_1 \to x_2 \to \cdots \to x_{k-1} \to x_k$$

be a **longest** path (of length $k - 1$) in T.

If $k = n$, then P is a Hamiltonian path in T, and we are through.

Suppose that $k < n$. We shall derive a contradiction.

Then, let v be a vertex in T but not in P.

As T is a tournament, either $v \to x_1$ or $x_1 \to v$. Since P is a longest path, $x_1 \to v$, otherwise $vx_1x_2 \cdots x_k$ will be a path longer than P. Again, either $v \to x_2$ or $x_2 \to v$. As P is longest, $x_2 \to v$, otherwise $x_1vx_2 \cdots x_k$ will be a path longer than P.

Proceeding in this manner successively, we arrive at the following, namely, $x_k \to v$. But then

$$x_1 \to x_2 \to \cdots \to x_{k-1} \to x_k \to v$$

is a path (of length k) in T longer than P, a contradiction. □

Remark. Theorem 7.2 says that every tournament always possesses a Hamiltonian path. A cycle in a tournament is called a **Hamiltonian cycle** if the cycle contains all vertices of the tournament. As 'Hamiltonian cycle' and 'Hamiltonian path' are so close, having Theorem 7.2, we may 'greedily' ask whether every tournament of order at least three contains a Hamiltonian cycle.

No! We are asking too much! Do not forget that Theorem 7.1 tells us that every transitive tournament contains no cycles, not to mention Hamiltonian ones.

Even if a tournament is non-transitive, it may contain no Hamiltonian cycles. For instance, the tournament of Figure 7.6 is non-transitive and it does not contain any Hamiltonian cycle.

So, which tournaments do contain Hamiltonian cycles?

Camion [C] proved in 1959 that *a tournament contains a Hamiltonian cycle if and only if it is a strong tournament.*

Theorem 7.2 provides us with an interesting property of tournaments. In what follows, we present another one.

Given a tournament T, no matter how large it is, by Theorem 7.2, we are guaranteed the existence of a vertex, from which we can reach all other vertices in T by a long (in fact, longest) but single path.

In contrast with this, we may ask a similar question, but from another extreme perspective, that is: Given any arbitrary tournament T, does there exist in T a vertex, from which we can reach all other vertices in T, each by some 'short' path?

What do 'short paths' mean? The shortest ones are, of course, of length one. In this case, the question is: Does there exist a vertex in T from which we can reach all others by paths of length one (that is, single arcs)?

As shown in Figure 7.7, clearly, this is possible when and only when T contains a source.

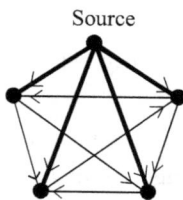

Figure 7.7

Now, how 'short' are the paths we can expect if T contains no source?

Naturally, we ask: *Does there exist a vertex in T from which we can reach all others by paths of length at most two?*

Example 7.4.1. *Let T be the tournament of Figure 7.8. Note that (i) T contains no source and (ii) from the vertex u, we can reach all others by paths of length at most two as indicated in bold.*

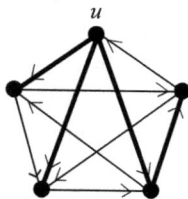

Figure 7.8

Question 7.4.3. *For each of the tournaments shown in Figure 7.6, Example 7.3.1 and Question 7.3.2, check whether there exists therein a vertex, from which we can reach all others by paths of length at most two.*

Vertices of this sort are interesting and we give them a nice name.

> A vertex w in a digraph is called a **king** if $d(w, x) \leq 2$ for every vertex x in D.

The existence of a king in any tournament is guaranteed by the following result obtained by Landau (see [L] and also Maurer [M]).

Theorem 7.3. *Let T be a tournament. If w is a vertex in T with maximum out-degree, then w is a king in T.*

Proof. Let x be an arbitrary vertex in T and $x \neq w$. Our aim is to show that $d(w, x) \leq 2$. If $w \to x$, then $d(w, x) = 1$, and we are through.

Assume now that $w \leftarrow x$. Let $O(w)$ be the set of vertices adjacent from w. If x dominates all vertices in $O(w)$ (see Figure 7.9(a)), then, together with the fact that $x \to w$, we have:

$$od(x) \geq |O(w)| + 1,$$

that is,

$$od(x) \geq od(w) + 1,$$

contradicting the maximality of $od(w)$.

Thus, x does not dominate all vertices in $O(w)$; that is, there exists a vertex in $O(w)$, say z, such that $z \to x$ (see Figure 7.9(b)).

Now, we have: $w \to z \to x$, and so $d(w, x) = 2$, as required. $\qquad \square$

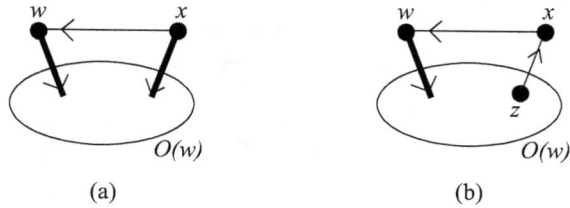

Figure 7.9

Question 7.4.4.   *Let T be a tournament and w a vertex in T. Theorem 7.3 says that if $od(w)$ is maximum, then w is a king in T. Is the converse true? That is, if w is a king in T, must $od(w)$ be maximum?*

Question 7.4.5.   *Construct a tournament of order 4 which contains exactly one king.*

Question 7.4.6.   *Construct a tournament of order 5 in which every vertex is a king.*

Exercise 7.4

(1) (+) Prove that a tournament is transitive if and only if it has one and only one Hamiltonian path.

(2) (+) Let T be a tournament and u, v be two vertices in T. If $od(u) \geq od(v)$, what are the possible values of $d(u, v)$? Justify your answer.

(3) (+) Suppose in a round-robin tournament, team A has the maximum score. Let p denote the number of teams defeated by A, and q the number of teams not defeated by A. Which of the following situations are possible?

   (i) $p > q$;
   (ii) $p = q$;
   (iii) $p < q$.

(4) For each $n \geq 2$, construct a tournament of order n in which there is a king w with $od(w) = 1$.

(5) (+) Let T be a tournament of order $n \geq 3$ and let u be a vertex in T with $od(u) \leq n - 2$. Show that u is dominated by a king in T.

(6) If w is the **only** king in a tournament of order n, what is the value of $od(w)$?

(7) (+) Is there a tournament which contains exactly two kings? Justify your answer.

(8) Suppose that a tournament T has exactly three kings. What can be said about the dominance relations among them?

(9) Using the results in Problem 18 of Exercise 7.2, show that for each $n \geq 3$, $n \neq 4$, there is a tournament of order n in which every vertex is a king.

(10) Show that there is no tournament of order four in which every vertex is a king.

(11) Let T be a regular tournament (see Problem 6 of Exercise 7.3). Is it true that every vertex in T is a king? Why?

(12) (+) A tournament T is said to be **irreducible** if T is **not** reducible, that is, for any partition of $V(T)$ into two non-empty subsets U and W, there exist $u \in U$, $w \in W$ such that $u \to w$ and there exist $y \in W$, $x \in U$ such that $y \to x$. (See Problem 16 of Exercise 7.3.)

(i) Is the following tournament irreducible?

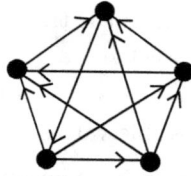

(ii) Prove that a tournament is strong if and only if it is irreducible.

(13) (+) Let T be a tournament. Prove that if T is strong, then every vertex in T is contained in a cycle. Is the converse true?

(14) (+) Let T be a strong tournament of order $n \geq 3$. Determine whether each of the following statements is true.

  (i) Every vertex in T is contained in a 3-cycle.
  (ii) Every arc in T is contained in a 3-cycle.
  (iii) Every arc in T is contained in a Hamiltonian cycle.
  (iv) Every arc in T is contained in a cycle.
  (v) For any two vertices u, v in T, either there is Hamiltonian path from u to v, or there is Hamiltonian path from v to u.

Chapter 8

Solutions of selected questions

8.1 Selected questions in Chapter 1

Exercise 1.2

Problem 10 *Solution.* (i) Vertex i represents soldier i. Two vertices are joined by an edge if the two corresponding soldiers can cooperate with each other and are not of the same specialization.

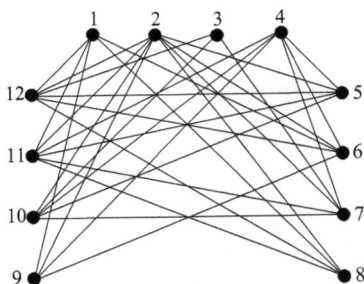

(ii) From the graph, each 3-man team $\{a, b, c\}$ satisfies the condition that each specialization is represented and all members of the term can work together if and only if the vertices corresponding to a, b, c are pairwise adjacent. Thus, the following sets provide such four 3-man teams:

$$\{1, 6, 9\}, \{2, 8, 12\}, \{3, 7, 10\}, \{4, 5, 11\}.$$

□

Exercise 1.3

Problem 4   Answer: G has 3 vertices of degree 5.

Problem 5   Answer: 20.

Problem 6   Answer: G has 4 vertices of degree 5.

Problem 11   Answer: $(n_2, n_3, n_4) = (2, 2, 4)$ or $(n_2, n_3, n_4) = (1, 4, 3)$.

Problem 15   Answer: $(k, n) = (1, 20), (2, 10)$ or $(4, 5)$.

Problem 16   Answer: $(n, m) = (6, 9)$. There are only two such G as shown below.

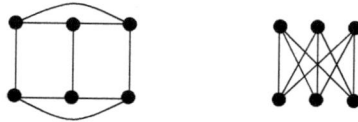

□

Problem 17   Answer: $15 \le n \le 20$.

Problem 20   *Solution.*   Model the situation as a graph G of order n, where the vertices are the persons, and two vertices are adjacent if and only if the two corresponding persons shook hands. By assumption, G is a simple graph. The problem is equivalent to showing that there exist two vertices u, v in G such that $d(u) = d(v)$.

It is clear that $0 \le d(x) \le n - 1$ for each vertex x in G. If the above statement is false, then there exist two vertices y and z in G such that $d(y) = 0$ and $d(z) = n - 1$, which however is impossible.   □

Problem 22   *Solution.*   Model the situation by a graph G with 8 vertices for 8 persons, and defining 'adjacency' for 'handshaking'. By assumption, $0 \le d(v) \le 6$ for each v in G, and each of '0, 1, 2, 3, 4, 5, 6' is the *degree* of some vertex.

Let v_1 be such that $d(v_1) = 6$ and $N(v_1) = \{v_2, v_3, \ldots, v_7\}$, say. Then $d(v_8) = 0$, and v_1 and v_8 are spouses (see the following).

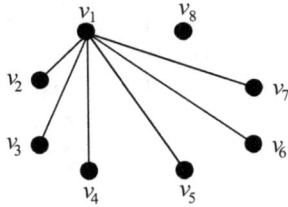

Let v_2 be such that $d(v_2) = 5$ and $N(v_2) = \{v_1, v_3, v_4, v_5, v_6\}$, say. Then $d(v_7) = 1$ and v_2 and v_7 are spouses (see below).

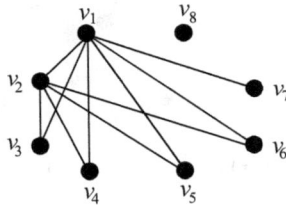

Let v_3 be such that $d(v_3) = 4$ and $N(v_3) = \{v_1, v_2, v_4, v_5\}$, say. Then $d(v_6) = 2$ and v_3 and v_6 are spouses (see below).

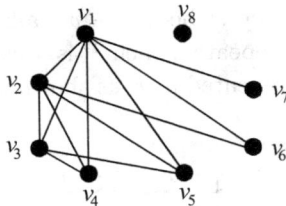

It follows that $d(v_4) = d(v_5) = 3$, and v_4 and v_5 are spouses.

As Mr Samy received different answers, either v_4 or v_5 represents Mrs Samy. Thus Mrs Samy shook hands with three others. □

Problem 23   *Solution.* Using a similar argument as shown in the solution of Problem 22, it can be shown that the answer is '$n - 1$' for this general problem. □

Problem 24   *Solution.*   Let $p_1, p_2, \ldots, p_n$ be the n given points in the plane. Form a graph G with $V(G) = \{p_1, p_2, \ldots, p_n\}$ in which two vertices are adjacent if their distance in the plane is '1'. What is the largest possible value that each $d(p_i)$ can have? By the assumption that the distance between any 2 points is at least one, it follows that $d(p_i) \leq 6$ (see the figure below).

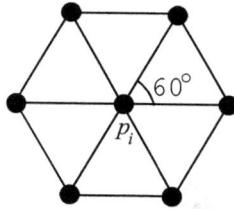

Thus, by Theorem 1.1,

$$2e(G) = \sum_{i=1}^{n} d(p_i) \leq 6n,$$

and so $e(G) \leq 3n$, as was to be shown.   □

Exercise 1.4

Problem 5   *Solution.*   Let P be a $u - v$ walk. We may assume that $u \neq v$. If no vertex in P is repeated, then P is a path, and we are through. Assume that a vertex x is repeated in P as shown below (it is possible that $x = u$ or $x = v$):

$$P : \underbrace{u \cdots}_{(a)} \underbrace{x \cdots}_{(b)} \underbrace{x \cdots v}_{(c)}.$$

Then P can be cut short by deleting the section (b) internally resulting in a shorter $u - v$ walk P' as shown below:

$$P' : \underbrace{u \cdots}_{(a)} \underbrace{x \cdots v}_{(c)}.$$

This procedure is repeatedly applied until no vertex in the resulting $u - v$ walk is repeated, and in this case, the resulting $u - v$ walk is a desired path.   □

Problem 6 *Solution.* Let Q be a circuit of length at least 2. If no vertex in Q is repeated, then Q is a cycle, and we are through. Assume that a vertex x is repeated in Q as shown below:

$$P : \underbrace{u \cdots}_{(a)} \underbrace{x \cdots}_{(b)} \underbrace{x \cdots u}_{(c)}.$$

Then Q can be cut short by deleting the section (b) internally resulting in a shorter circuit Q' as shown below:

$$Q' : \underbrace{u \cdots}_{(a)} x \underbrace{\cdots u}_{(c)}.$$

This procedure is repeatedly applied until no vertex in the resulting circuit is repeated, and in this case, the resulting circuit is a desired cycle.

$\square$

Problem 7 Hint: Consider a longest path $P = v_0 v_1 \cdots v_r$ in G.

Problem 8 Hint: Prove by contradiction and suppose $d(u, v) > 2$ holds for some pair of vertices u and v in G.

Problem 9 Hint: Show that $e(G) \leq \binom{n-1}{2}$ if G is disconnected.

Problem 12 Hint: Apply the inclusion-exclusion principle.

Problem 18 *Solution.* Let P and Q be two longest paths (of length k each, say) in a connected graph G, and suppose on the contrary that they have no vertex in common. As G is connected, there exist a vertex u in P and a vertex v in Q which are joined by a path R, say. Without loss of generality, we may assume (see the figure below) that (i) this $u - v$ path R contains no vertex in P or Q other than u and v, (ii) the length of the subpath (a) in P is greater than or equal to that of (b) and (iii) the length of the subpath (d) in Q is greater than or equal to that of (c).

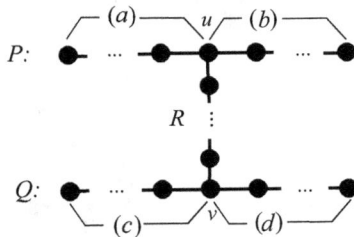

With this, however, we observe that the path consisting of the subpath (a) in P, the $u - v$ path R and the subpath (d) in Q is of length greater than k, a contradiction.   □

Problem 20

Proof.   Starting at an arbitrary vertex, say v_0, in G, we walk along distinct edges in G to produce a maximal trail (no edge is repeated), say of length s, and label the edges along the trail $1, 2, \ldots, s$ (see the example below).

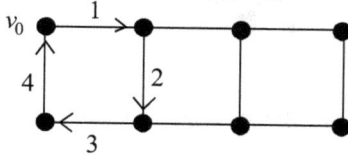

If there are edges not yet labeled, as G is connected, one of them is incident with a vertex, say v_r, which has been visited. Starting at v_r, we walk along distinct unlabelled edges in G to produce another maximal trail, say of length p, and label the edges along the trail $s + 1, s + 2, \ldots, s + p$ (see the diagram below).

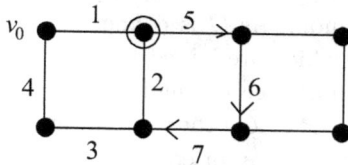

We repeat the above procedure until all edges in G are labeled (see the diagram below).

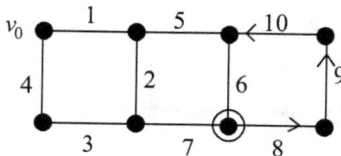

We now show that for each vertex v with $d(v) \geq 2$ in G, the *gcd* of the labels of the edges incident with v is 1. If $v = v_0$, the situation is clear as the first edge incident with it is labeled 1. Assume that $v \neq v_0$. Let e be the edge with which we first visit v via a trail. As $d(v) \geq 2$, along the same trail, we leave v with a new edge, say f. By the above procedure, the labels of e and f are consecutive numbers, say t and $t + 1$, and so the corresponding *gcd* (that is, $gcd\{t, t + 1, \ldots\}$) is 1. $\square$

8.2   Selected questions in Chapter 2

Exercise 2.2

Problem 6   *Solution.*   (i) $G \cong G$ since the identity mapping is one-to-one and onto, and clearly preserves adjacency.

(ii) Suppose $G \cong H$. Then there exists a one-to-one and onto mapping f from $V(G)$ to $V(H)$ which preserves adjacency. Clearly, f^{-1} is a one-to-one and onto mapping from $V(H)$ to $V(G)$ which preserves adjacency. Thus, $H \cong G$.

(iii) Suppose $G \cong H$ and $H \cong J$. Then there exist one-to-one and onto mappings, f from $V(G)$ to $V(H)$ and g from $V(H)$ to $V(J)$, which preserve adjacency. Now, the composite mapping $g \circ f$ is a one-to-one and onto mapping from $V(G)$ to $V(J)$ which preserves adjacency. Thus, $G \cong J$.
                                                                              $\square$

Exercise 2.3

Problem 19   *Solution.*   Yes, G must contain a bridge in this case. The justification is given below.

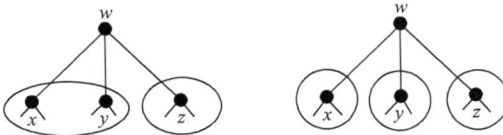

Let w be a cut-vertex in G. As G is cubic (i.e. 3-regular), w has exactly

3 neighbors, say x, y and z (see the diagrams below). Observe that $G - w$ is disconnected and one of its components contains exactly one of the x, y and z, say z. In this case, wz is a bridge in G.   □

Problem 24   *Solution.*   Start by picking a vertex at random and naming it x_1. Pick any unnamed neighbor of x_1, and name it x_2. In general, having named vertices with the names $x_1, x_2, \ldots, x_k$, check through all neighbors of x_k. If there is such a vertex unnamed, pick one and name it x_{k+1}. Otherwise, find the largest index j such that x_j has an unnamed neighbor. Pick such an unnamed neighbor, and name it x_{k+1}.

An example of the above procedure is shown below. For convenience, we denote x_j by 'j'.

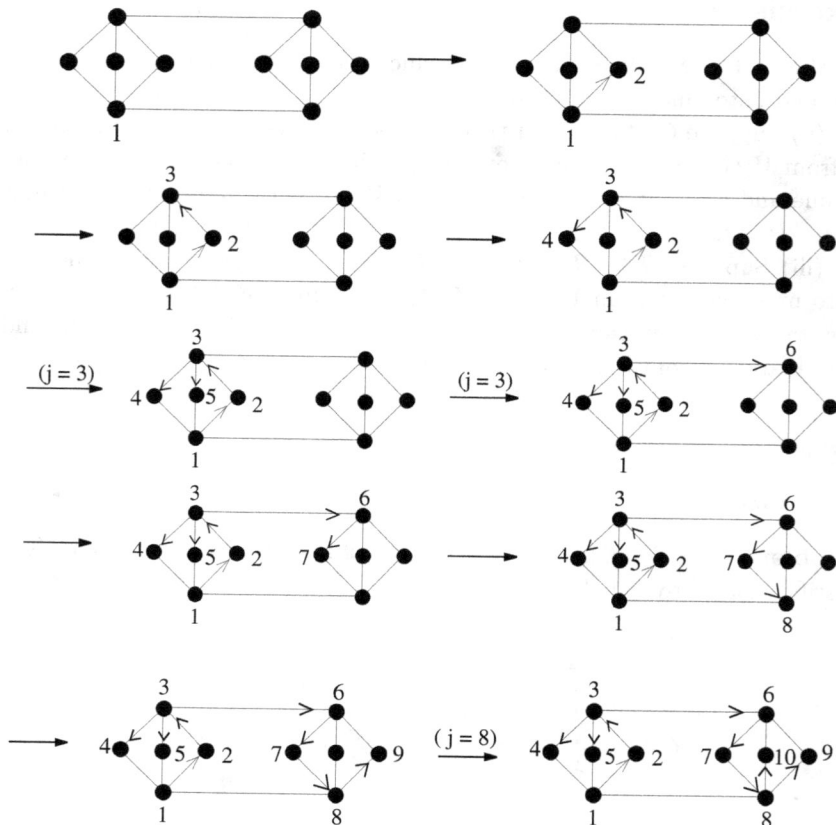

We shall now prove that the subgraph $[\{x_1, x_2, \ldots, x_i\}]$ is connected for $i = 1, 2, \ldots, n$ by induction. The statement is obvious for $i = 1, 2$. Assume that it is true for $i = k$. Consider the case that $i = k + 1$. We know from the procedure that the vertex x_{k+1} is a neighbor of either x_k or x_j, where $1 \leq j \leq k$. As $[\{x_1, x_2, \ldots, x_k\}]$ is connected by the induction hypothesis, it follows that $[\{x_1, x_2, \ldots, x_{k+1}\}]$ is also connected.

This completes the proof.

$\square$

Problem 29   *Solution.*

(i) For $A = \{u, v, z\}$, $e(A, V(H)\backslash A) = 7$.

(ii) We first note that for $A \subseteq V(G)$, the following equality holds (see the diagram below):

$$\sum_{x \in A} d(x) = e(A, V(G)\backslash A) + \sum_{x \in A} d_{[A]}(x) \tag{1}$$

where $d_{[A]}(x)$ denotes the degree of x in $[A]$.

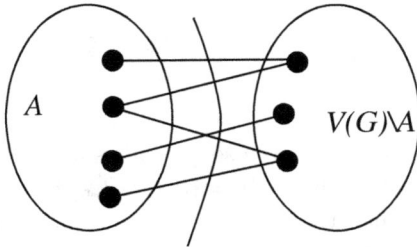

Let A_e and A_o be, respectively, the set of even vertices and the set of odd vertices in A. Then

$$\sum_{x \in A} d(x) = \sum_{x \in A_e} d(x) + \sum_{x \in A_o} d(x).$$

By Theorem 1.1, the sum $\sum_{x \in A} d_{[A]}(x)$ is always even. It thus follows from

(1) that

$$e(A, V(G)\backslash A) \text{ is even} \iff \sum_{x \in A} d(x) \text{ is even}$$

$$\iff \sum_{x \in A_e} d(x) + \sum_{x \in A_o} d(x) \text{ is even}$$

$$\iff \sum_{x \in A_o} d(x) \text{ is even}$$

$$\iff |A_o| \text{ is even}.$$

The proof is thus complete.   □

Exercise 2.4

Problem 16   *Solution.*   A regular self-complementary graph of order 9 is shown below:

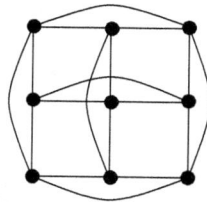

□

Problem 19   *Solution.*   (i) We use 'graph' as a model to study the problem. Let G be a graph with $V(G) = \{a, b, c, d, e, f\}$, which represents the group of six people, such that two vertices are *adjacent* in G if and only if the two corresponding people are mutual acquaintances. Thus, two vertices are *non-adjacent* in G if and only if the two corresponding people are complete strangers to one another. Our objective is to show that either G contains a triangle or $\overline{G}$ contains a triangle.

Consider the vertex a. Either $d(a) \geq 3$ in G or $d_{\overline{G}}(a) \geq 3$. Since $\overline{\overline{G}} \cong G$, we may assume that $d(a) \geq 3$ in G. Let b, c, d be in $N(a)$. If any two in $\{b, c, d\}$ are adjacent, say b and c, then $[\{a, b, c\}]$ is a triangle in G, and we are through. Otherwise, no two in $\{b, c, d\}$ are adjacent in G, which means that $[\{a, b, c\}]$ forms a triangle in $\overline{G}$.

This completes the proof.

(ii) The result in (i) is no longer true for five vertices. Consider the cycle C_5. Both C_5 and $\overline{C_5}$ ($\cong C_5$) contain no triangles. $\square$

Problem 20 *Solution.* Let G be a graph of order 6 satisfying the condition

(*) *containing no N_3 as an induced subgraph.*

By Problem 19 above, G contains either a triangle or a N_3 as an induced subgraph. Since the latter cannot happen by (*), G contains at least one triangle. We claim that G contains at least 'two' triangles. Suppose on the contrary that G contains exactly one triangle, say $xyzx$, as shown below:

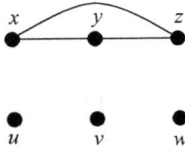

(1) Consider $\{x, u, v, w\}$. We assert that x must be adjacent to one of u, v and w. If not, then by applying (*) to $\{x, u, v\}$, $\{x, u, w\}$ and $\{x, v, w\}$, it follows that $[\{u, v, w\}]$ forms a triangle, a contradiction.

Thus, say, x is adjacent to u as shown below:

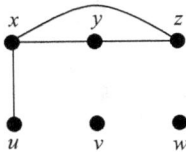

(2) Consider $\{y, u, v, w\}$. Likewise, y must be adjacent to one of u, v and w. To avoid producing another triangle, y must be adjacent to one of v and w, say v as shown in the following:

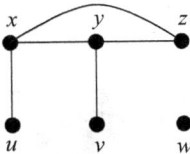

(3) Consider $\{z, u, v, w\}$. Likewise, z must be adjacent to one of u, v and

w. To avoid producing another triangle, z must be adjacent to w as shown below:

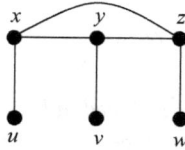

(4) Apply (*) to $\{x, v, w\}$. To avoid producing another triangle, v and w must be adjacent as shown below:

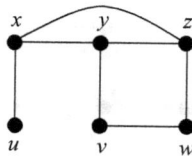

(5) Apply (*) to $\{y, u, w\}$. To avoid producing another triangle, u and w must be adjacent as shown below:

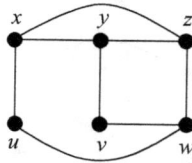

(6) Consider $\{z, u, v\}$. To avoid producing another triangle, no two in $\{z, u, v\}$ are adjacent. But then $[\{z, u, v\}] \cong N_3$, contradicting (*).

We conclude from the above discussion that G contains at least two triangles. The following graph satisfies (*) and contains exactly two triangles. Thus, the least possible value for $n_G(C_3)$ is '2'.

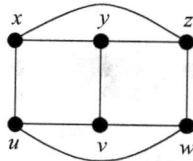

□

Problem 22   *Solution.*   Let G be a graph of odd order and $\delta(G) \geq 5$. By Corollary 1.2, not all vertices in G are of degree 5. Thus, there is a vertex, say w, in G with $d(w) \geq 6$. Consider $N(w)$. By Problem 19(i), $[N(w)]$ contains either a triangle or an induced N_3. The latter cannot happen by assumption. Thus, $[N(w)]$ contains a triangle. It follows that $[N(w) \cup \{w\}]$, and hence G, contains a K_4.

Remark. Indeed, by Problem 20, G contains at least two K_4's.   □

Problem 23   *Solution.*   (i) Let G be a graph of order 9 which contains no triangles.

Case (1). $\Delta(G) \geq 4$. Let w be a vertex in G such that $d(w) \geq 4$ and let a, b, c, d be adjacent to w in G. As G contains no triangles, no two in $\{a, b, c, d\}$ can be adjacent in G. Thus $[\{a, b, c, d\}]$ forms a K_4 in $\overline{G}$.

Case (2). $\Delta(G) \leq 3$. Then $\delta(\overline{G}) \geq 5$. As G contains no triangles, $\overline{G}$ contains no N_3 as an induced subgraph. Thus $\overline{G}$ contains a K_4 by the result in Problem 22.

(ii) The conclusion in (i) is no longer true if $n = 8$. Consider the following graph G of order 8. It can be checked that G contains no triangles and $\overline{G}$ contains no K_4 as well.

G:

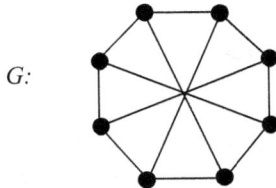

□

Problem 24   *Solution.*   Before we present a proof for this problem, let us revisit Problem 19(i), which says that for any graph G of order 6, either G or $\overline{G}$ contains a triangle. By superimposing $\overline{G}$ onto G so that the same vertices are identified, we obtain a K_6. Thus the problem can equivalently be stated as: Coloring the edges of K_6 by 'blue' (for G) or 'red' (for $\overline{G}$), either there is a 'blue triangle' or a 'red triangle' in K_6.

We shall now generalize the above idea to solve the problem. Consider K_{17} in which each vertex represents a person. Color the edges in K_{17} by three colors as follows: an edge uv is colored blue (respectively, red and yellow) if u and v discuss topic I (respectively, II and III). Our aim is to show that there is a 'blue triangle', a 'red triangle' or a 'yellow triangle' in K_{17}.

Let w be a vertex in K_{17}. As the 16 edges incident with w are colored by three colors, by the Pigeonhole Principle, at least 6 of the edges are colored by one same color, say blue. Let wa, wb, wc, wd, we and wf be any six of such blue edges.

Now consider the $K_6 = [\{a, b, c, d, e, f\}]$. If one of the edges in this K_6, say ab, is colored blue, then we have a blue triangle, namely, $wabw$.

If none of the edges in this K_6 is colored blue, then all the edges in this K_6 are colored red or yellow, and so there must be a red triangle or a yellow triangle in this K_6 by the result in Problem 19(i).

The proof is thus complete.   □

8.3   Selected questions in Chapter 3

Exercise 3.1

Problem 8   Hint: Apply Result (1) in Section 3.1 and the fact that 5 is the only number in the degree sequence of G which is not a multiple of 3.

Problem 9   *Solution.* Let X be the set of boys and Y the set of girls at a party. Let G be a bipartite graph in which b ($\in X$) and g ($\in Y$) are joined by an edge if and only if b dances with g. Our aim is to show that there exist b, b' in X and g, g' in Y such that the adjacency relations shown in the following diagram holds, where "●⋯⋯●" indicates the non-adjacency of the two vertices.

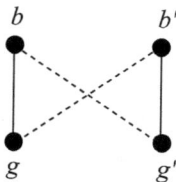

Let b be a vertex in X such that $d(b) = \max\{d(x) \mid x \in X\}$. Clearly, $d(b) \geq 1$; i.e. $N(b)$ is non-empty. As no boy dances with every girl, there exists a vertex, say g', in $Y \backslash N(b)$. Since each girl dances with at least one boy, there exists a vertex, say b', in X such that b' and g' are adjacent. Clearly, $b' \neq b$. Now, if b' is adjacent to every vertex in $N(b)$, then $d(b') \geq d(b) + 1$, contradicting the maximality of $d(b)$ in X. Thus, there exists a vertex, say g, in $N(b)$ which is not adjacent to b'. It follows that the four vertices chosen, namely b, b', g and g', are the desired ones (see the diagram below).

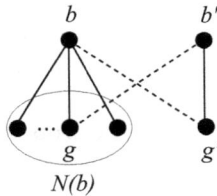

$$N(b)$$

□

Problem 10   Hint: Apply Result (1) in Section 3.1 and the given conditions.

Problem 11   Hint: Apply Result (1) in Section 3.1 and the given conditions.

Problem 21   *Solution.*   (i) There are three induced cycles, namely, $xywx$, $yzwy$ and $abcwya$.

(ii) ($\Rightarrow$) Suppose on the contrary that G contains an induced cycle C of odd order. Then C is itself an odd cycle in G, and so G is not bipartite by Theorem 3.1, a contradiction.

($\Leftarrow$) Suppose on the contrary that G is not bipartite. Then, by Theorem 3.1, G contains an odd cycle C. If C is induced, then we are through; otherwise, there are two vertices, say u and v, in C, which are not adjacent along C but are adjacent in G (see the diagram below).

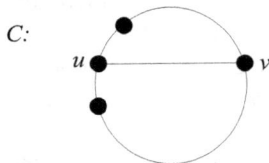

C:

As C is odd, one of the $u - v$ paths along C forms, together with the edge uv, a smaller odd cycle C'. If C' is induced, then we are through; otherwise, the above argument = can be repeated in a finite number of steps to eventually reveal an induced cycle of odd order in G.   □

Problem 22   *Solution.*   The graph H must be bipartite. We prove it by contradiction as follows.

Suppose on the contrary that H is not bipartite. Then, by Theorem 3.1, H contains an odd cycle $C : v_1v_2 \cdots v_{2k+1}v_1$. We may assume that v_1 is odd. But then, by assumption, v_2 is even, which, in turn, implies eventually that v_{2k+1} is odd. This, however, contradicts the assumption as these two odd vertices v_1 and v_{2k+1} are joined an edge.

The graphs $K(p, q)$, where p and q are of different parity, are examples.
Note. Not every bipartite graph has the property described in the problem.
                    □

Problem 25   *Solution.*   ($\Rightarrow$) If $G \cong K(p, q)$, where $2 \leq p \leq q$, then it is clear that every two edges in G are contained in a common C_4.

($\Leftarrow$) Let (X, Y) be a bipartition of G.
Claim: G is a complete bipartite graph.

If not, then there exist u in X and v in Y which are not adjacent in G. As $\delta(G) \geq 1$, assume that u is adjacent to y in Y and v is adjacent to x in X (see the diagram below).

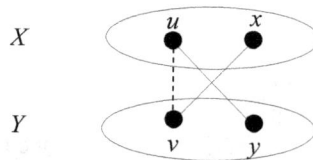

By assumption, the edges uy and vx are contained in a common C_4. This, however, implies that uv is an edge, a contradiction. Thus, G is a complete bipartite graph with bipartition (X, Y).

We shall now show that $G \cong K(p, q)$. Let $|X| = k$, where $1 \leq k \leq \frac{p+q}{2}$. Then $|Y| = p + q - k$ and $pq = e(G) = k(p + q - k)$. The latter implies that $(k - p)(k - q) = 0$. Since $p \leq q$ and $k \leq \frac{p+q}{2}$, we have $k = p$.

Now, as $k = p$, $|Y| = q$, and we have $G \cong K(p, q)$.   □

Exercise 3.2

Problem 8   *Solution.*   Let G be a unicyclic graph of order n and let n_i denote the number of vertices of degree i in G, $i = 1, 2, \ldots, k(= \Delta(G))$. Then, by Problem 7(i), $e(G) = v(G) = n$, and by Theorem 1.1,

$$\sum_{v \in V(G)} d(v) = 2e(G) = 2n.$$

But

$$\sum_{v \in V(G)} d(v) = n_1 + 2n_2 + \cdots + kn_k.$$

Thus

$$n_1 + 2n_2 + \cdots + kn_k = 2n = 2(n_1 + n_2 + \cdots + n_k).$$

and we have

$$n_1 = n_3 + 2n_4 + \cdots + (k-2)n_k.$$

□

Problem 10   *Solution.*   The proof is by induction on k. For $k = 1, 2$, the result is trivial.

Assume that the result is true when $k = n$. We now consider the case when $k = n+1$. Thus, let T be a tree of order $n+1$ and G be a graph with $\delta(G) \geq n$. We shall show that T is isomorphic to some subgraph of G. Let w be an end-vertex of T and suppose that w is adjacent to u in T (see the diagram below). Write $T' = T - w$. Since $v(T') = n$ and $\delta(G) \geq n > n-1$, by the induction hypothesis, T' is isomorphic to some subgraph, say T'', of G.

Let u' be the image of u in G under an isomorphism. Observe that $d(u') \geq \delta(G) \geq n$ in G and $v(T'') = n$. Thus, u' is adjacent to a vertex in G, say v, which is not in T''. Clearly, the subgraph of G, which consists of T'' and the edge $u'v$, is isomorphic to T. The proof is thus complete.   □

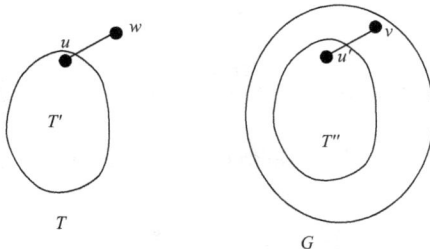

Problem 14   Hint: Apply the Handshaking Lemma and the fact that $\sum\limits_{v\in V(T)} d(v) = n_1 + 2n_2 + \cdots + kn_k.$.

Problem 15   Hint: Apply the facts that each tree with at least two vertices contains one vertex of degree 1, and removing any vertex of degree 1 from a tree always yields a smaller tree.

Problem 17   *Solution.*   We shall prove the statement by induction on $n(\geq 2)$.

For $n = 2$, as $d_1 + d_2 = 2$, $d_1 = d_2 = 1$; and the sequence $(1,1)$ is the degree sequence of the path of order 2.

Assume that the statement is true for all sequences of length $n - 1$, where $n \geq 3$.

Consider now the sequence $(d_1, d_2, \ldots, d_n)$ with

$$\sum_{i=1}^{n} d_i = 2(n-1).$$

We may assume that $d_1 \geq d_2 \geq \cdots \geq d_n$.

Clearly, $d_n = 1$; otherwise, $d_1 \geq d_2 \geq \cdots \geq d_n \geq 2$, and we have

$$\sum_{i=1}^{n} d_i \geq 2n,$$

a contradiction.

As $n \geq 3$, let k be the largest index in $\{1, 2, \ldots, n-1\}$ such that $d_k \geq 2$. Let $d'_k = d_k - 1$. Then

$$(d_1, d_2, \ldots, d_{k-1}, d'_k, \ldots, d_{n-1}) \tag{1}$$

is a sequence of positive integers of length $n - 1$ such that

$$\begin{aligned} &d_1 + d_2 + \cdots + d_{k-1} + d'_k + \cdots + d_{n-1} \\ &= 2(n-1) - 2 \qquad\qquad (d_n = 1 \text{ and } d'_k = d_k - 1) \\ &= 2((n-1) - 1). \end{aligned}$$

By the induction hypothesis, the sequence (1) is the degree sequence of some tree T of order $n - 1$.

Let v be a vertex in T of degree d'_k. Construct a tree T^* of order n by adding to T a new vertex w and joining w to v (thus $d(w) = 1$ and $d(v) = d_k$ in T^*). Clearly, $(d_1, d_2, \ldots, d_n)$ is the degree sequence of T^*.

The proof is thus complete.   □

Problem 19 *Solution.* Let G be a graph of order n and size $n-1$.

($\Rightarrow$) Assume that G is connected. Since $e(G) = v(G) - 1$, by Theorem 3.4, G is a tree, and so contains no cycles.

($\Leftarrow$) Assume that G contains no cycles. Then G is a forest (see Problem 18). By the result of Problem 18 (iii), $c(G) = v(G) - e(G)$. Thus, by the given assumption, $c(G) = n - (n-1) = 1$, which means that G is connected. □

Exercise 3.3

Problem 6 *Solution.* Let G be a connected bipartite graph with bipartition (X, Y). By Result (1) in Section 3.1,

$$e(G) = \sum_{x \in X} d(x).$$

As $d(x) \le 7$ for each x in X, $e(G) \le 7|X|$.

On the other hand, as G is connected, by Corollary 3.7,

$$e(G) \ge v(G) - 1 = |X| + |Y| - 1.$$

Combining the above two inequalities, we have: $|X| + |Y| - 1 \le e(G) \le 7|X|$; that is, $|Y| \le 6|X| + 1$.

An example of such a connected bipartite graph with $|Y| = 6|X| + 1$ is shown below:

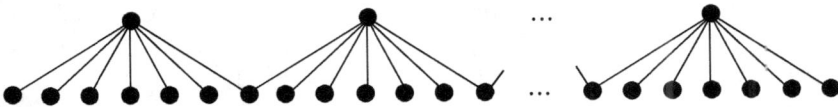

□

Problem 7 *Solution.* Let G be a connected bipartite graph with bipartition (X, Y). By Result (1) in Section 3.1,

$$e(G) = \sum_{x \in X} d(x).$$

As $d(x) \le 4$ for each x in X,

$$e(G) = \sum_{x \in X} d(x) \le 4|X|.$$

On the other hand, as G is connected but not a tree, $e(G) > v(G) - 1$ by Corollary 3.7 and Theorem 3.4.

Combining the above two inequalities, we have:
$$|X| + |Y| = v(G) \leq e(G) \leq 4|X|;$$
that is, $|Y| \leq 3|X|$. The following example shows that $|Y| = 3|X|$ holds when $|X| = 2$.

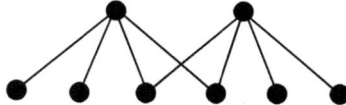

Note. The reader is encouraged to construct such G for which $|Y| = 3|X|$ holds for $|X| = 3, 4, \ldots$. $\square$

8.4   Selected questions in Chapter 4

Exercise 4.3

Problem 7   *Solution.*   Let G be a graph of order $n \geq 2$. We claim that $\chi(G) = n - 1$ if and only if $G \not\cong K_n$ and G contains a K_{n-1} as a subgraph.

Suppose $G \not\cong K_n$ and G contains a K_{n-1}. Then from the result that $\chi(G) = n$ if and only if $G \cong K_n$, we have $\chi(G) < n$. Also, G contains a K_{n-1} implies that $\chi(G) \geq n - 1$. Thus $\chi(G) = n - 1$.

Suppose $\chi(G) = n - 1$. From the result that $\chi(G) = n$ if and only if $G \cong K_n$, we have that $G \not\cong K_n$. We shall show that K_{n-1} is a subgraph of G.

Suppose on the contrary that K_{n-1} is not a subgraph of G. Then $G - x \not\cong K_{n-1}$ for every $x \in V(G)$.

Since $G \not\cong K_n$, there exist $u, v \in V(G)$ such that $uv \notin E(G)$. Then $xy \in E(G)$ for any $x, y \in V(G) \setminus \{u, v\}$; otherwise, $[\{u, v, x, y\}]$ is 2-colourable, implying that $\chi(G) \leq n - 2$, a contradiction. Hence $G - \{u, v\} \cong K_{n-2}$.

Note that $G - u \not\cong K_{n-1}$ and $G - v \not\cong K_{n-1}$. Since $G - \{u, v\} \cong K_{n-2}$, there exist $u', v' \in V(G) \setminus \{u, v\}$ such that $uu', vv' \notin E(G)$.

If $u' = v'$, then u, v and u' can be assigned the same colour, implying that $\chi(G) \leq n - 2$, a contradiction.

If $u' \neq v'$, then the four vertices u, u', v, v' can be coloured by two colours, implying that $\chi(G) \leq n - 2$, a contradiction too.

Therefore K_{n-1} is a subgraph of G. □

Note. The following is another proof that *if G is of order n and $\chi(G) = n - 1$, then G contains K_{n-1} as a subgraph.*

Let $V(G) = \{v_1, v_2, \ldots, v_n\}$ and θ be a $(n-1)$-colouring of G. We may assume that $\theta(v_i) = i$ for $i \leq n - 1$, $\theta(v_n) = 1$ and $d(v_n) \leq d(v_1)$. Consider the subgraph H induced by $V(G) \setminus \{v_n\}$. Suppose $H \ncong K_{n-1}$. We shall show that G can be recoloured with a $(n-2)$-colouring. We have that $v_i v_j \notin E(H)$ for some $i < j \leq n - 1$. If $i \neq 1$, we may recolour v_i as j thus obtaining a $(n-2)$-colouring of G (with colour i excluded), a contradiction. If $i = 1$, we may recolour v_1 as j. Note that $d(v_n) \leq d(v_1) \leq n - 3$. Thus, v_n can be recoloured with at least one colour from $2, 3, \ldots, n - 1$. Thus, G admits a $(n-2)$-colouring (with colour 1 excluded), which is a contradiction. Hence, $H \cong K_{n-1}$ and so G contains a K_{n-1} as a subgraph. □

Problem 31   *Solution.*   For the solutions below, unless otherwise stated, let V_i be the set of vertices v in G with $\theta(v) = i$ for each $i = 1, 2, \ldots, \chi(G)$, where θ is a $\chi(G)$-colouring of the graph G.

(i) The set of 4 'white' vertices indicated in the figure below is an independent set in H. It is easy to check that there is no independent set in H with 5 vertices. Thus, $\alpha(H) = 4$.

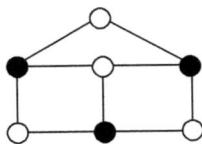

(ii) Since $\alpha(G) \geq \max\{|V_i| \mid i = 1, 2, \ldots, \chi(G)\}$, we have

$$\chi(G)\alpha(G) \geq \sum_{i=1}^{\chi(G)} |V_i|,$$

i.e., $\chi(G)\alpha(G) \geq n$.

(iii) The following connected graph H is such that $v(H) = 12$, $\chi(H) = 4$ and $\alpha(H) = 3$.

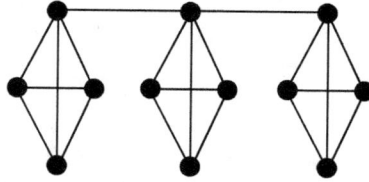

(iv) Let A be an independent set in G such that $|A| = \alpha(G)$. Then the number of vertices in $G - A$ is $n - \alpha(G)$. Introduce a colouring of G as follows: all vertices in A are coloured by one colour, and each of the $n - \alpha(G)$ vertices in $G - A$ is coloured by one new colour. Clearly, this defines a $(n + 1 - \alpha(G))$-colouring of G. Thus, $\chi(G) \leq n + 1 - \alpha(G)$ and so $\chi(G) + \alpha(G) \leq n + 1$.

(v) $K_4 + N_7$ (see below) is a connected graph with the required properties.

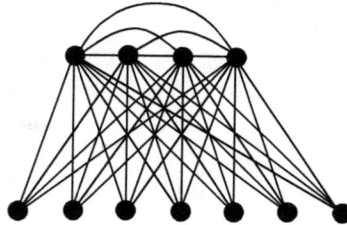

$\square$

Problem 32 *Solution.* Since G is not bipartite, $\chi(G) \geq 3$. By assumption, $G - S$ does not contain an odd cycle. Therefore $G - S$ is bipartite and is 2-colourable. It is possible to colour all vertices in S with one colour because S is independent. Thus G has a 3-colouring. Hence, $\chi(G) = 3$. $\square$

Problem 33 *Solution.* (i) Let $G = \bigcup_{i=1}^{k} H_i$, where $k \geq 2$ and the H_i's are the components of G. Then $\chi(G) = \max\{\chi(H_i) | i = 1, 2, \ldots, k\}$. We may assume that $\chi(G) = \chi(H_1) = 5$. Then, $\chi(G - v) = \chi(H_1) = 5$ for each vertex v in H_2, a contradiction. Thus, G is connected.

(ii) Suppose $\delta(G) \leq 3$. Let v be a vertex in G such that $d(v) = \delta(G)$.

Since $\chi(G - v) = 4$, there is a colouring $\theta : V(G - v) \rightarrow \{1, 2, 3, 4\}$. Since $d(v) = \delta(G) \leq 3$, there is at least one colour, say 4, not used to colour the neighbours of v. Extend θ to colour G by colouring v the colour 4. This gives a 4-colouring of G. Thus, $\chi(G) \leq 4$, a contradiction. Hence $\delta(G) \geq 4$.

(iii) Suppose there exist two vertices u, v in G such that $N(u)$ is a subset of $N(v)$ (then u and v are non-adjacent). Since $\chi(G - u) = 4$, there is a 4-colouring θ of $G - u$. Since $N(u)$ is a subset of $N(v)$, θ can be extended to a 4-colouring of G by colouring u the colour $\theta(v)$. Thus $\chi(G) = 4$, contradicting the condition $\chi(G) = 5$. Hence $N(u)$ is not a subset of $N(v)$ for any two vertices u, v in G.

(iv) Suppose $v(G) = 6$. From Problem 7, $G \neq K_6$ and G contains a K_5. Then there exists a vertex v in G such that $G - v = K_5$. However, this means that $\chi(G - v) = 5$, a contradiction. Thus, $v(G) \neq 6$. $\qquad\square$

Exercise 4.4

Problem 8   (i) Answer: 3.

Exercise 4.5

Problem 5   *Solution.*   No, such a G does not exist. We shall prove it by contradiction.

Suppose such a G exists. Let H be a component of G with $\chi(H) = \chi(G) = 7$. Clearly, H is not an odd cycle. If H is a complete graph, then by (ii), $H \cong K_r$, where $r \leq 6$. But then $\chi(H) = \chi(K_r) = r \leq 6$, a contradiction. Thus, H is a connected graph which is neither an odd cycle nor a complete graph. By Brooks' Theorem, $\chi(H) \leq \Delta(H) \leq 6$, a contradiction again. Thus, no such G exists. $\qquad\square$

Problem 6   *Solution.*   As G is connected, $\chi(G) \geq 2$. By Theorem 4.1, $\chi(G) \leq \Delta(G) + 1 = 3 + 1 = 4$. Thus, $2 \leq \chi(G) \leq 4$.

If $G \cong K_4$ (3-regular), then $\chi(G) = 4$. If $G \not\cong K_4$, then G can never be complete. As G is cubic, G is not an odd cycle. Hence, by Brook's Theorem, $\chi(G) \leq \Delta(G) = 3$. We thus have the following conclusion:

$$\chi(G) = \begin{cases} 4 & \text{if } G \cong K_4; \\ 2 & \text{if } G \text{ is bipartite}; \\ 3 & \text{otherwise.} \end{cases}$$

$\qquad\square$

Problem 7   *Solution.*   ($\Leftarrow$) Suppose $G \cong K_n$. Then $\overline{G} \cong N_n$. Thus, $\chi(G) + \chi(\overline{G}) = n + 1$. Suppose $G \cong C_5$. Then $\overline{G} \cong C_5$. In this case, $\chi(G) + \chi(\overline{G}) = 3 + 3 = 5 + 1$.

($\Rightarrow$) Assume that $\chi(G) + \chi(\overline{G}) = n + 1$ and suppose $G \not\cong K_n$. We shall prove that $G \cong C_5$.

Let G be k-regular. Then $2 \leq k \leq n - 2$ and $\overline{G}$ is $(n - 1 - k)$-regular. Assume that G is not an odd cycle. Then by Brook's Theorem, $\chi(G) \leq \Delta(G) = k$. As $\chi(\overline{G}) \leq \Delta(\overline{G}) + 1 = (n - 1 - k) + 1 = n - k$ by Theorem 4.1, we have

$$\chi(G) + \chi(\overline{G}) \leq k + (n - k) = n,$$

a contradiction. Thus $G \cong C_{2r+1}$ (here, $n = 2r + 1$). If $r = 1$, then $G \cong K_3$, which is not allowed. Assume $r \geq 3$. Then $\overline{G}\ (= \overline{C_{2r+1}})$ is a connected graph which is neither complete nor an odd cycle. By Brook's Theorem, $\chi(\overline{G}) \leq \Delta(\overline{G}) = (n - 1) - 2 = n - 3$. Thus,

$$\chi(G) + \chi(\overline{G}) \leq 3 + (n - 3) = n,$$

a contradiction.

Hence $r = 2$, and so $G \cong C_5$, as required. $\square$

Exercise 4.6

Problem 4   *Solution.*   Let Z be the graph with 7 vertices which represent the 7 subjects A, B, C, D, E, F and G, where two vertices are adjacent if and only if there is a pupil taking the two subjects represented by the vertices.

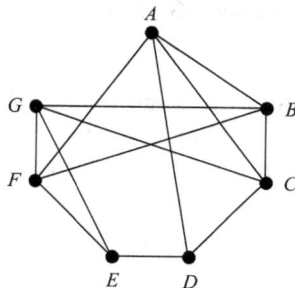

Since Z contains a triangle, $\chi(Z) \geq 3$. Suppose $\chi(Z) = 3$. Let $\theta :$ $V(Z) \to \{1, 2, 3\}$ be a 3-colouring of Z. We may assume that the vertices

A and B are coloured '1' and '2' respectively. Then C and F must both be coloured '3'. Next, G and E must be coloured '1' and '2' in turn. However, D now cannot be coloured with '1', '2' or '3', a contradiction. Thus, $\chi(Z) \geq 4$. The figure below shows a 4-colouring of Z. Thus, the minimum number of timeslots needed is 4 and a suitable time allocation of the subjects is $\{A, G\}$, $\{B, E\}$, $\{C, F\}$ and $\{D\}$.

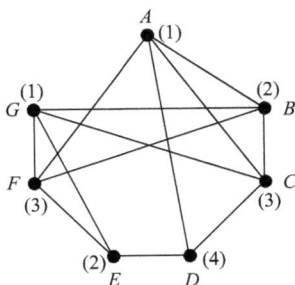

8.5 Selected questions in Chapter 5

Exercise 5.2

Problem 1 *Solution.* (a) The bipartite graph is shown below:

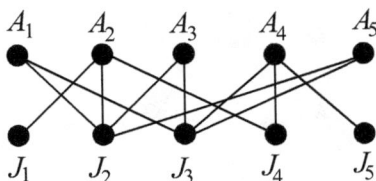

(b) It is impossible. If it is possible to assign each applicant to a job for which he/she applies, then all the five jobs must be assigned to the five applicants. Thus J_1 must be assigned to A_2 and J_5 must be assigned to A_4. But, then J_4 cannot be assigned to any one, a contradiction. □

Problem 2 (b) Answer: It is possible.

Problem 4 (b)   Answer: Yes.

(c) Hint: Suppose M_1 marries W_1. Draw a new bipartite graph to model the new situation.

Problem 5   Hint: Construct a bipartite graph such that the partite sets are the set of codewords $\{ab, abc, cd, bcd, de\}$ and the set of letters $\{a, b, c, d, e\}$, and a codeword is adjacent only to the letters contained in it.

Problem 7   Answer: (i) Yes. (ii) No (iii) Yes. (iv) Yes. Yes. (v) No. Hint: (vi) By (v), G has no perfect matching containing by. So we just need to consider perfect matchings in $G - by$.

Problem 8   *Solution.*   A perfect matching of C_n has exactly $n/2$ edges. Thus, if n is odd, there are no perfect matchings.

Suppose n is even. Let C_n be $v_1 v_2 \cdots v_n v_1$ and let M be a perfect matching of C_n. If $v_1 v_2$ is in M, then $v_i v_{i+1}$, for all odd i, is in M. On the other hand, if $v_1 v_2$ is not in M, then $v_i v_{i+1}$, for all even i, is in M. Thus, there are exactly two perfect matchings for C_n when n is even.   □

Problem 9   Answer: $n!$.

Problem 10   Answer: $(n-1)(n-3)\cdots 3 \cdot 1$ when n is even.

Problem 11   Answer: (i) No. (ii) No. (iv) 12.

Problem 12   *Solution.*   Colour the 34 squares black and white as shown below:

Clearly, each domino covers two squares with different colours. Form a bipartite graph G with bipartition (X, Y), where X is the set of black squares and Y is the set of white squares, and a vertex in X is adjacent to a vertex in Y if and only if their corresponding squares can be covered by a domino. The problem is equivalent to asking whether G contains a perfect matching. The answer is 'no' as $|X| = 18$ and $|Y| = 16$.   □

Problem 13 Hint: Let us label the vertices of the triple flyswat G as follows:

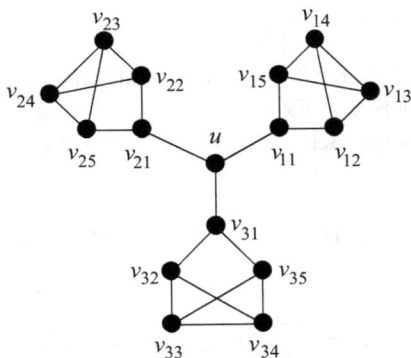

Suppose the graph G has a perfect matching M. We may assume that $uv_{11} \in M$.

Problem 14 Answer: (i) $\lfloor \frac{n}{2} \rfloor$, (ii) 1.

Problem 15 Hint: Represent the situation as a graph G with 20 vertices representing the 20 players and two vertices are adjacent if and only if a game is scheduled between the two corresponding players. Thus, there are 14 edges and the degree of each vertex is at least 1.

Let M be a matching in G such that $|M|$ is the largest among all the matchings in G. Let U be the set of vertices in G which are not incident with any edge in M. Then $|U| = 20 - 2|M|$, as shown below.

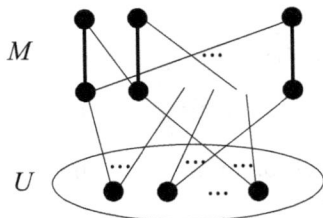

Exercise 5.3

Problem 1 *Solution.* We find that

$$N(S) = \{J_2, J_3\}.$$

So $|N(S)| = 2 < 3 = |S|$. By Theorem 5.1, the bipartite graph that models the situation in Problem 1 of Exercise 5.2 has no complete matchings. Thus it is impossible to assign each applicant a job for which he/she applies. $\square$

Problem 5 Answer: It is true in general that $E_1 \subseteq E_2$.

Problem 6 *Solution.* (i) Since $d(x) \geq k$ for each $x \in X$, we have

$$|E_1| = \sum_{x \in S} d(x) \geq \sum_{x \in S} k = k|S|.$$

Since $d(y) \leq k$ for each $y \in Y$, we have

$$|E_2| = \sum_{y \in N(S)} d(y) \leq \sum_{y \in N(S)} k = k|N(S)|.$$

By the result of Problem 5, $|E_1| \leq |E_2|$. Thus we have

$$k|S| \leq |E_1| \leq |E_2| \leq k|N(S)|.$$

(ii) By (i), we have $|S| \leq |N(S)|$ for every subset S of X. Thus, by Theorem 5.1, G has a complete matching from X to Y.

(iii) Let G be a k-regular bipartite graph with bipartition (X, Y), where $k \geq 1$. By the result of Problem 4 in Exercise 3.1, we have $|X| = |Y|$. Thus, by (ii), G has a perfect matching. $\square$

Problem 7 Hint: Let H be the subgraph of G induced by $X^* \cup N(X^*)$.

Problem 8 Hint: Use Theorem 5.1 for both directions of the proof.

Problem 9 Hint: Use Theorem 5.1 for both directions of the proof.

Problem 10 *Solution.* (i) $\Rightarrow$ (ii)

Since G contains perfect matchings, we have $|X| = |Y|$, and by Theorem 5.1, $|S| \leq |N(S)|$ for all $S \subset X$.

Suppose that there exists $S \subset X$ with $S \neq \emptyset$ such that $|S| = |N(S)|$.

Since G is connected, there exists an edge xy with $x \in X \backslash S$ and $y \in N(S)$.

By (i), xy is contained in a perfect matching M of G. But then the vertices in S are matched with the vertices in $N(S) \setminus \{y\}$ under M, which is impossible as $|S| > |N(S) \setminus \{y\}|$.

(ii) $\Rightarrow$ (iii)

Let $x \in X$ and $y \in Y$ and $H = G - \{x, y\}$. Note that $(X \setminus \{x\}, Y \setminus \{y\})$ is a bipartition of H.

Let S be any set with $S \neq \emptyset$ and $S \subseteq X \setminus \{x\}$. We have $|N_F(S)| \geq |N_G(S)| - 1$. As $S \subset X$, by (ii), we have $|N_G(S)| \geq |S| + 1$. Thus

$$|N_H(S)| \geq |N_G(S)| - 1 \geq |S| + 1 - 1 = |S|.$$

By Theorem 5.1, H has a complete matching from $X \setminus \{x\}$ to $Y \setminus \{y\}$. Since $|X| = |Y|$ by (ii), this complete matching is perfect.

(iii) $\Rightarrow$ (i)

Let xy be any edge in G, where $x \in X$ and $y \in Y$.

By (iii), $G - \{x, y\}$ has a perfect matching M. Clearly, $M \cup \{xy\}$ is a perfect matching of G. $\square$

Problem 11   *Solution.*   ($\Rightarrow$) Suppose that G^* has a perfect matching M. Let

$$M' = M \setminus \{e \in M \mid e \text{ is incident with a vertex } y \in Y^*\}.$$

Then M' is a matching of G with $|M'| = |M| - |Y^*| = |X| - p = |Y|$.

($\Leftarrow$) Assume that G has a matching M' with $|Y|$ edges.

Let $X' = \{x \in X \mid x \text{ is not incident with any edge in } M'\}$. Then $|X'| = p$ and $[X' \cup Y^*] \cong K(p, p)$. Clearly, $[X' \cup Y^*]$ possesses a perfect matching, say M''. Then $M' \cup M''$ is a perfect matching of G^*. $\square$

Problem 12

Solution.   By Theorem 5.1, the result holds if $k = |X|$. In the following we assume that $1 \leq k < |X|$.

Construct the bipartite graph G^* with bipartition $(X, Y \cup Y^*)$ such that G is an induced subgraph of G^* and

(1) $|Y^*| = |X| - k$;

(2) every vertex in Y^* is adjacent to every vertex in X.

($\Rightarrow$) Assume that G contains a matching M with $|M| = k$.

Let

$$X' = \{x \in X \mid x \text{ is not incident with any edge in } M\}.$$

Clearly, $|X'| = |X| - k = |Y^*|$, and by the definition of G^*, $[X' \cup Y^*]$ is a complete bipartite graph with bipartition (X', Y^*). Thus $[X' \cup Y^*]$ contains a perfect matching, say M'.

It follows that $M \cup M'$ is a complete matching of G^* from X to $Y \cup Y^*$. By Theorem 5.1, for all $S \subseteq X$,

$$|S| \le |N_{G^*}(S)| = |N_G(S)| + |Y^*| = |N(S)| + |X| - k.$$

($\Leftarrow$) Assume that for all $S \subseteq X$,

$$|S| \le |N(S)| + |X| - k.$$

As $N_{G^*}(S) = N(S) \cup Y^*$, we have

$$|N_{G^*}(S)| = |N(S)| + |Y^*| = |N(S)| + |X| - k.$$

Thus, for all $S \subseteq X$,

$$|S| \le |N_{G^*}(S)|.$$

By Theorem 5.1, G^* has a complete matching M^* from X to $Y \cup Y^*$. Let

$$M = M^* \backslash \{e \in M^* \mid e \text{ is incident with a vertex in } Y^*\}.$$

Then M is a matching of G with

$$|M| = |M^*| - |Y^*| = |X| - (|X| - k) = k.$$

This completes the proof. $\square$

Problem 13 *Solution.* By the result of Problem 12, we need only to show that for all $S \subseteq X$,

$$|S| \le |N(S)| + |X| - \left\lceil \frac{3}{4}|X| \right\rceil.$$

If $|S| \le |N(S)| + |X| - \frac{3}{4}|X|$, then

$$|S| \le \left\lfloor |N(S)| + |X| - \frac{3}{4}|X| \right\rfloor = |N(S)| + |X| - \left\lceil \frac{3}{4}|X| \right\rceil.$$

Note that $|N(S)| + |X| - \frac{3}{4}|X| = |N(S)| + \frac{1}{4}|X|$. So it suffices to show that

$$|S| \le |N(S)| + \frac{1}{4}|X|$$

for all $S \subseteq X$.

Let S be any subset of X and $H = [S \cup N(S)]$. Then $(S, N(S))$ is a bipartition of H. Note that

$$e(H) = \sum_{x \in S} d_H(x) = \sum_{x \in S} d_G(x) \ge 6|S|$$

and

$$e(H) = \sum_{y \in N(S)} d_H(y) \le \sum_{y \in N(S)} d_G(y) \le 8|N(S)|.$$

Thus

$$6|S| \le 8|N(S)|,$$

i.e., $\frac{3}{4}|S| \le |N(S)|$. Now, we have

$$|S| = \frac{3}{4}|S| + \frac{1}{4}|S| \le \frac{3}{4}|S| + \frac{1}{4}|X| \le |N(S)| + \frac{1}{4}|X|.$$

By the result of Problem 11, G contains a matching M with $|M| \ge \frac{3}{4}|X|$.
$\square$

Problem 14 *Solution.* Let S be any subset of X and $H = [S \cup N(S)]$. Then $(S, N(S))$ is a bipartition of H. Note that

$$e(H) = \sum_{x \in S} d_H(x) = \sum_{x \in S} d_G(x) \ge 3|S|$$

and

$$e(H) = \sum_{y \in N(S)} d_H(y) \le \sum_{y \in N(S)} d_G(y) \le 4|N(S)|.$$

Thus

$$4|N(S)| \ge 3|S|,$$

i.e., $|N(S)| \ge \frac{3}{4}|S|$, which implies that

$$\rho(S) = |S| - |N(S)| \le |S| - \frac{3}{4}|S| = \frac{1}{4}|S| \le \frac{1}{4}|X|.$$

It follows that $|X| \ge 4\rho(S)$ for all $S \subseteq X$.
$\square$

Problem 15 *Solution.* (a)

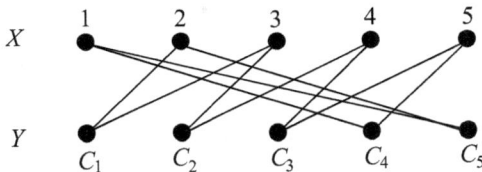

(b) $d(i) = 2$ for each $i \in X$, as every i does not appear in exactly 2 columns.

(c) $d(C_j) = 2$ for each $C_j \in Y$, as each column excludes exactly 2 members in X.

(d) Yes, G is 2-regular.

(e) Yes, G contains a perfect matching. By Corollary 5.3, every k-regular bipartite graph with $k \geq 1$ contains a perfect matching.

(f) The following is a perfect matching of G:

$$\{1C_5, 2C_1, 3C_2, 4C_3, 5C_4\}.$$

(g) From the above perfect matching, we get the following 4×5 Latin rectangle L':

$$L' = \begin{pmatrix} 1\ 2\ 3\ 4\ 5 \\ 5\ 1\ 2\ 3\ 4 \\ 4\ 5\ 1\ 2\ 3 \\ 2\ 3\ 4\ 5\ 1 \end{pmatrix}.$$

(h) We expand L' to form the following Latin square:

$$L'' = \begin{pmatrix} 1\ 2\ 3\ 4\ 5 \\ 5\ 1\ 2\ 3\ 4 \\ 4\ 5\ 1\ 2\ 3 \\ 2\ 3\ 4\ 5\ 1 \\ 3\ 4\ 5\ 1\ 2 \end{pmatrix}.$$

□

Problem 16 *Solution.* (a)

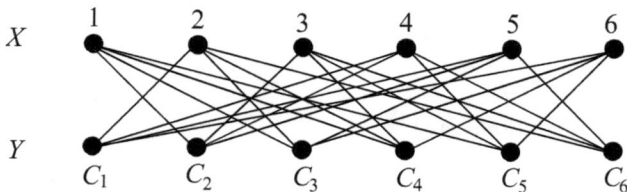

(b) $d(i) = 4$ for each $i \in X$, as every i does not appear in exactly 4 columns.

(c) $d(C_j) = 4$ for each $C_j \in Y$, as each column excludes exactly 4 numbers in X.

(d) Yes, G is 4-regular.

(e) Yes, G contains a perfect matching. Since G is 4-regular, by Corollary 5.3, it contains a perfect matching.

(f) The following is a perfect matching of G:

$$\{1C_4, 2C_6, 3C_2, 4C_5, 5C_3, 6C_1\}.$$

(g) From the above perfect matching, we get the following 3×6 Latin rectangle L':

$$L' = \begin{pmatrix} 1\ 2\ 3\ 4\ 5\ 6 \\ 3\ 6\ 4\ 5\ 2\ 1 \\ 6\ 3\ 5\ 1\ 4\ 2 \end{pmatrix}.$$

□

Exercise 5.4

Problem 3 *Solution.* (i) The graph G is shown below:

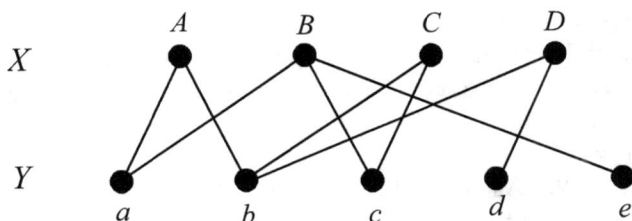

(ii) If $S = \{A\}$, then $N(S) = \{a, b\}$.

(iii) If $S = \{A, B\}$, then $N(S) = \{a, b, c, e\}$.

(iv) If $S = \{A, B, C\}$, then $N(S) = \{a, b, c, e\}$.

(v) There are no subsets S of X such that $|S| > |N(S)|$.

(vi) By Theorem 5.4, there exists a complete matching from X to Y.

(vii) The following is a complete matching M from X to Y:

$$M = \{Aa, Bc, Cb, Dd\}.$$

(viii) From M given in (vii), (A, B, C, D) has an SDR: (a, c, b, d). □

Problem 7 *Solution.* (i) Note that $(1, 2, 3, \cdots, n)$ is an SDR of $(S_1, S_2, \cdots, S_n)$.

(ii) We claim that $(S_1, S_2, \cdots, S_n)$ has only one SDR, i.e., $(1, 2, 3, \cdots, n)$. Suppose that $(S_1, S_2, \cdots, S_n)$ has an SDR $(x_1, x_2, \cdots, x_n)$.

Then $x_1 \in S_1$ and so $x_1 = 1$; $x_2 \in S_2 \setminus \{1\}$ and so $x_2 = 2$. Likewise, we have $x_i = i$ for each $i = 3, \cdots, n$. Thus $(x_1, x_2, \cdots, x_n) = (1, 2, \cdots, n)$. $\square$

Problem 8 *Solution.* (i) Since $i \in S_i$ for $i = 1, 2, \cdots n$, we have

$$\left| \bigcup_{i \in I} S_i \right| \geq |I|$$

for all $I \subseteq \{1, 2, \cdots, n\}$. By Theorem 5.4, $(S_1, S_2, \cdots, S_n)$ has an SDR.

(ii) We shall show that $(S_1, S_2, \cdots, S_n)$ has exactly two SDR's, namely,

$$(1, 2, \cdots, n) \quad \text{and} \quad (2, 3, \cdots, n, 1).$$

Let $(x_1, x_2, \cdots, x_n)$ be an SDR of $(S_1, S_2, \cdots, S_n)$.

Since $S_1 = \{1, 2\}$, we have either $x_1 = 1$ or $x_1 = 2$.

Case 1: $x_1 = 1$.

Then $x_n \neq 1$. As $S_n = \{n, 1\}$, we have $x_n = n$. Then $x_{n-1} \neq n$. As $S_{n-1} = \{n-1, n\}$, we have $x_{n-1} = n-1$. Continuing this argument, we have $x_i = i$ for each $i = n-2, n-3, \cdots, 2$. Thus $(x_1, x_2, \cdots, x_n) = (1, 2, \cdots, n)$.

Case 2: $x_1 = 2$.

Then $x_2 \neq 2$. As $S_2 = \{2, 3\}$, $x_2 = 3$. Then $x_3 \neq 3$. As $S_3 = \{3, 4\}$, $x_3 = 4$. Continuing this argument, we have $x_i = i + 1$ for each $i = 2, 3, \cdots, n-1$ and $x_n = 1$. Thus $(x_1, x_2, \cdots, x_n) = (2, 3, \cdots, n, 1)$. $\square$

Problem 9 *Solution.* If $a = 1$, then $S_1 = \{1\} = S_2$, and so (S_1, S_2, S_3, S_4) has no SDR.

If $a = 2$, then $S_1 = \{1, 2\}$, $S_2 = \{1, 3\}$, $S_3 = \{2\}$ and $S_4 = \{2, 3\}$. Since

$$|S_1 \cup S_2 \cup S_3 \cup S_4| = 3 < 4,$$

(S_1, S_2, S_3, S_4) has no SDR.

If $a = 3$, then $S_1 = \{1, 3\}$, $S_2 = \{1, 5\}$, $S_3 = \{2, 1\}$ and $S_4 = \{2, 4\}$. Observe that (S_1, S_2, S_3, S_4) has an SDR: $(1, 5, 2, 4)$.

If $a = 4$, then $S_1 = \{1, 4\}$, $S_2 = \{1, 7\}$, $S_3 = \{2, 0\}$ and $S_4 = \{2, 5\}$. Observe that (S_1, S_2, S_3, S_4) has an SDR: $(1, 7, 2, 5)$.

Hence (S_1, S_2, S_3, S_4) has an SDR if and only if $a \in \{3, 4\}$. $\square$

Problem 10 *Solution.* Let $X = \{S_1, S_2, \cdots, S_{12}\}$ be the set of 12 clubs and Y the set of students at the college. Let G be the bipartite graph with bipartition (X, Y) such that $S_i(\in X)$ is adjacent to $y(\in Y)$ if and only if $y \in S_i$.

By the given conditions, we have

$$d(y) \leq 3 \leq d(S_i)$$

for each $y \in Y$ and $i = 1, 2, \cdots, 12$.

Thus, by the result of Problem 6 (ii) in Exercise 5.3, G contains a complete matching M from X to Y. Let $M = \{S_1y_1, S_2y_2, \cdots, S_{12}y_{12}\}$. Then $(y_1, y_2, \cdots, y_{12})$ is an SDR of $(S_1, S_2, \cdots, S_{12})$. $\quad\square$

8.6 Selected questions in Chapter 6

Exercise 6.1

Problem 1 *Solution.* (i) $e(G) = 8$.

(ii) Name the vertices as shown below:

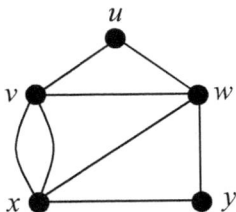

A circuit with 2 edges : xvx.

A circuit with 3 edges : $uvwu$.

A circuit with 4 edges : $xvwyx$.

(iii) The circuit $xvuwvx$ is an example.

(iv) The circuit $uvwxywu$ is an example.

(v) Eight ($= e(G)$) edges.

(vi) Yes, G contains an Euler circuit, e.g. $uvxvwxywu$. $\quad\square$

Problem 2   *Solution.*   For each of the five Eulerian multigraphs, an Euler circuit is exhibited by labelling its edges as shown below.   □

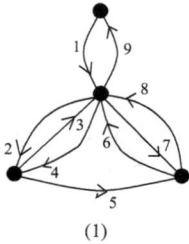

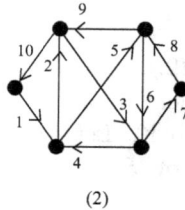

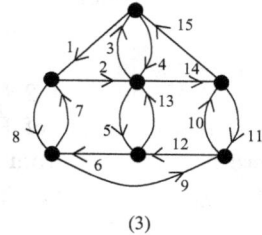

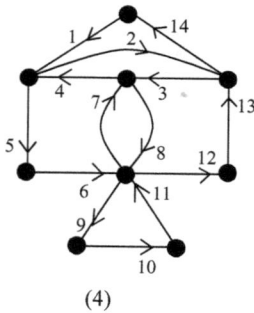

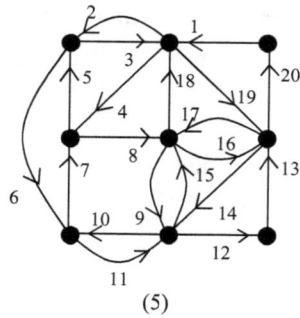

(1)   (2)   (3)

(4)   (5)

Exercise 6.2

Problem 7   *Solution.*   Let G be a connected multigraph in which every vertex is even. We shall prove that G is Eulerian by induction on $e(G)$ (≥ 2), the size of G.

When $e(G) = 2$, clearly,

$$G \; \cong \; \bullet \!\!\!\!\!\!\!\!\!\!\!\!\!\!\! \bullet$$

and when $e(G) = 3$,

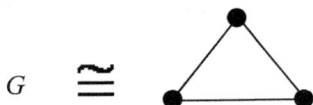

and they are Eulerian.

Assume that the statement is true for all connected multigraphs G with $e(G) \le k - 1$, where $k \ge 4$, in which every vertex is even.

Let G be a connected multigraph with $e(G) = k$ (≥ 4) in which every vertex is even. If G is 2-regular, then G is a cycle, and we are through.

Assume now that there is a vertex, say w, in G with $d(w) \ge 4$. Let e and f be two edges incident with w in G and let $H = G - \{e, f\}$.

Case 1. The edges e and f are parallel (see the diagram below).

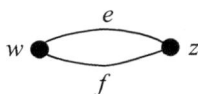

Clearly, every vertex in H is even (but H may not be connected).

(i) H is connected. As $e(H) < k$, by the induction hypothesis, H is Eulerian, and hence H possesses an Euler circuit W. Clearly, W can be extended to an Euler circuit W' of G by inserting e and f successively into W the moment the vertex w is visited when W is traversed.

(ii) H is disconnected. Then H has two components, say, H^* and H' (see the diagram below). By the induction hypothesis, both H^* and H' are Eulerian, and hence they possess Euler circuits W^* and W' respectively. Combine W^*, W' and $\{e, f\}$ as follows: starting at w and following W^* to return to w; traversing e to reach z and following W' to return to z; and finally, traversing f to terminate at w. This is clearly an Euler circuit of G.

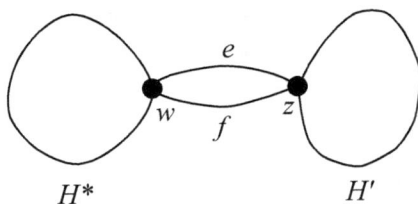

Case 2. The edges e and f are non-parallel (see the diagram below, where $e = wx$ and $f = wy$).

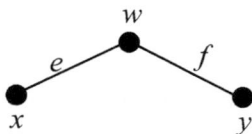

In this case, H has two odd vertices, namely, x and y. Let R be the multigraph obtained from H by adding a new edge joining x and y. Then every vertex in R is even (but R may not be connected).

(i) R is connected. As $e(R) < k$, by the induction hypothesis, R is Eulerian, and hence R possesses an Euler circuit W. Clearly, W can be converted to an Euler circuit of G if we replace the new edge joining x and y, say xy, in R by $e = xw$ and $f = wy$ successively.

(ii) R is disconnected. Then R has two components, say, R^* and R', where R^* contains w while R' contains the new edge xy (see the diagram below). By the induction hypothesis, both R^* and R' are Eulerian, and hence they possess Euler circuits W^* and W' respectively. We shall now combine $W^*, W' - xy$ and $\{e, f\}$ to produce an Euler circuit of G. We may assume that xy is the first edge in W'. Starting at x, we traverse $e(= xw)$ to reach w; following the whole W^* from w, we are back to w; traversing $f(= wy)$, we are now at y; finally, following the rest of $W' - xy$, we terminate at x.

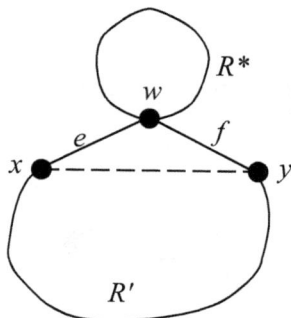

Clearly, the above closed walk is an Euler circuit of G. The proof is thus complete. □

Problem 8 *Solution.* ($\Rightarrow$) Assume that G is semi-Eulerian. By definition, G possesses an open Euler trail W, say from x to y. Clearly, the trail
W' obtained from W by adding yx at the end of W is an Euler circuit of
$G + xy$ (the multigraph obtained from G by adding xy). Thus, $G + xy$ is
Eulerian, and by Theorem 6.1, all vertices in $G + xy$ are even. It follows
that x and y are the only two odd vertices in G.

($\Leftarrow$) Assume that G has exactly two odd vertices, say x and y. Then
all vertices in $G + xy$ are even, and so $G + xy$ possesses an Euler circuit W.
We may assume that xy is the first edge in traversing W. Then $W - xy$
forms an Euler $y - x$ trail of G (see the solution of Problem 6(iii) in this
Exercise). This shows that G is semi-Eulerian. $\square$

Problem 10 *Solution.* ($\Rightarrow$) We shall prove that the edges in an Eulerian multigraph can always be partitioned into some edge-disjoint cycles by
induction on p, the number of cycles in G. It is obvious if $p = 1$. Assume
that it holds for Eulerian multigraphs with $p < k$, where $k \geq 2$. Let G be
an Eulerian multigraph with $p = k$. By Theorem 6.1, all vertices in G are
even. Choose a cycle, say C, in G. Let H be the multigraph obtained from
G by deleting the edges in C (note that H may be disconnected). Observe
that each component of H has all its vertices even (and hence is Eulerian),
and possesses less than k cycles. Thus, by the induction hypothesis, the
edges in each component of H can be partitioned into edge-disjoint cycles.
Accordingly, the edges in G can be partitioned into edge-disjoint cycles, in
which C is a member.

($\Leftarrow$) Let G be a connected multigraph. Assume now that the edges in
G can be partitioned into edge-disjoint cycles. We shall show that G is
Eulerian. Let C be a cycle in this partition. If C includes all edges in G,
then G is Eulerian. Otherwise, as G is connected, there is another cycle,
say C', in this partition which has at least one vertex v in common with C.
The (closed) walk P that starts at v and consists of the cycles C and C'
in succession is a closed trail containing the edges of these two cycles. If P
includes all edges of G, then G is Eulerian. Otherwise, applying a similar
argument, P can be extended to a longer closed trail P' of G. Continuing
this process, a closed trail containing all edges in G can eventually be
obtained, which then shows that G is Eulerian. $\square$

Problem 12   *Solution.*   There are $2q$ odd vertices in G_2. To turn them into even in the resulting multigraph, it requires at least $2q$ new edges. The following construction shows that $2q$ new edges suffice.   □

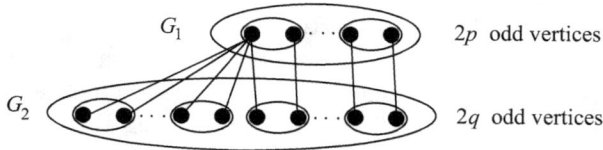

Problem 13   *Solution.*   There are 6 odd vertices as shown in H.

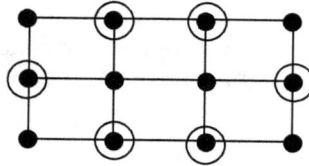

We pair them up into 3 pairs and join each pair by a new edge. Thus the least number of new edges is '3'. This can be done in $5 \times 3 \times 1$ ($= 15$) ways.   □

Problem 15   *Solution.*   ($\Leftarrow$) By Theorem 6.1, we show that each vertex in G is even. Thus, let v be in $V(G)$ and $A = \{v\}$. Observe that $d(v) = e(A, V(G)\backslash A)$, which is even by assumption.

($\Rightarrow$) [First proof]

Let A be a proper subset of $V(G)$ and W an Euler circuit of G with the starting vertex v in A. Clearly, for each edge e in W from A to $V(G)\backslash A$, there is an edge e' in W from $V(G)\backslash A$ to A (as v is also the terminal vertex of W). It follows that $e(A, V(G)\backslash A)$ is even.

($\Rightarrow$) [Second proof]

For $A \subseteq V(G)$, we have (see the solution of Problem 29 in Exercise 2.3):

$$e(A, V(G)\backslash A) = \sum_{x \in A} d(x) - \sum_{x \in A} d_{[A]}(x)$$

where $d_{[A]}(x)$ denotes the degree of x in $[A]$. By assumption and Theorem 6.1, $\sum_{x \in A} d(x)$ is even, and by Theorem 1.1, $\sum_{x \in A} d_{[A]}(x)$ is even. It follows from the above equality that $e(A, V(G)\backslash A)$ is even.   □

Problem 16 *Solution.* Let G be a graph which contains $K(5,6)$ as a spanning subgraph.

(i) If G is semi-Eulerian, then the minimum size of G is 32 ($= 5 \times 6 + 2$). An example of such G is shown below:

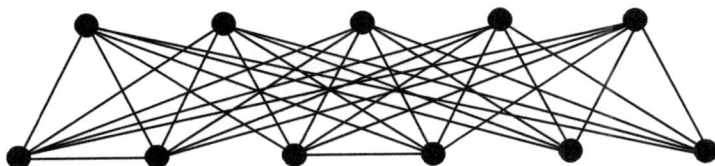

(ii) If G is Eulerian, then the minimum size of G is 33 ($= 5 \times 6 + 3$). An example of such G is shown below:

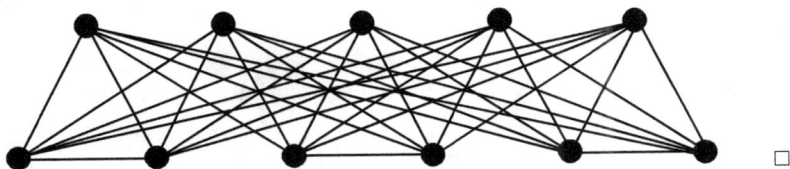

□

Problem 17 *Solution.* Let G be a multigraph which contains $K(5,7)$ as a spanning subgraph.

(i) Assume that G is semi-Eulerian.

Yes, G can be simple, and the minimum size of G is 40 ($= 5 \times 7 + 5$). An example of such G is shown below:

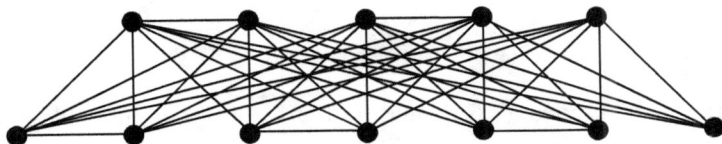

(ii) Assume that G is Eulerian. We first show that G cannot be simple. Suppose on the contrary that G is simple and let (X, Y) be the bipartition of $K(5, 7)$ with $|X| = 5$. Since $d(x)$ is even in G, $d_{[X]}(x)$ ($= d(x) - 7$) is odd in $[X]$ for each x in X. Thus the graph $[X]$ consists of 5 odd vertices, which contradicts Corollary 1.2.

The minimum size of G is 41 ($= 5 \times 7 + 6$). An example of such G is shown in the following: □

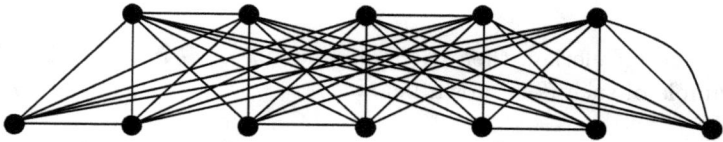

Problem 18   *Solution.*   Let G be an Eulerian graph of order 8 and size 10.

(i) Let k be the maximum possible value of $\Delta(G)$. As G is Eulerian, $\Delta(G)$ is even, and so $\Delta(G) \leq 6$. We claim that $k = 6$.

Let x, y and z be, respectively, the number of vertices of degree 2, 4 and 6 in G. Then

$$x + y + z = 8 \tag{8.6.1}$$

and

$$2x + 4y + 6z = 20,$$

or

$$x + 2y + 3z = 10 \tag{8.6.2}$$

Solving the equations (8.6.1) and (8.6.2) yields one possible solution $(x, y, z) = (7, 0, 1)$. It follows that $k = 6$.

There is only one such G as shown below:

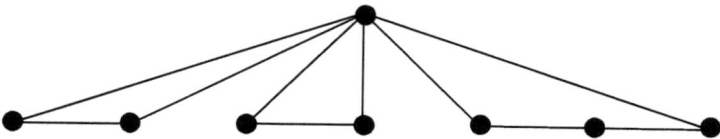

(ii) Suppose that $\Delta(G) = 4$.

(a) Then solving (1) and (2) in (a) with $z = 0$ gives $(x, y) = (6, 2)$. Thus there are 2 vertices of degree 4 in G.

(b) Let u and v be the two non-adjacent vertices of degree 4 in G. Note that

$$2 \leq |N(u) \cap N(v)| \leq 3.$$

Case 1. $|N(u) \cap N(v)| = 2$.

There are two such G's as shown below:

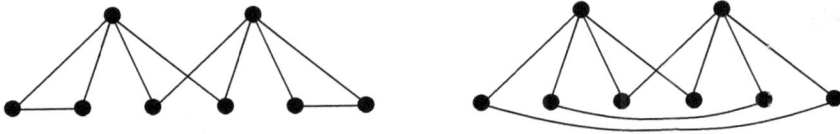

Case 2. $|N(u) \cap N(v)| = 3$.

There is only one such G as shown below:

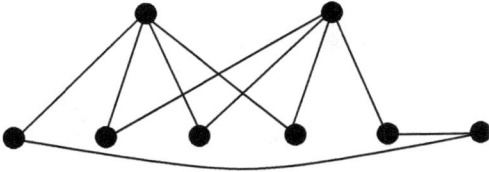

□

Exercise 6.4

Problem 3 *Solution.* ($\Leftarrow$) We may assume that n is even. In this case, a spanning cycle is shown below:

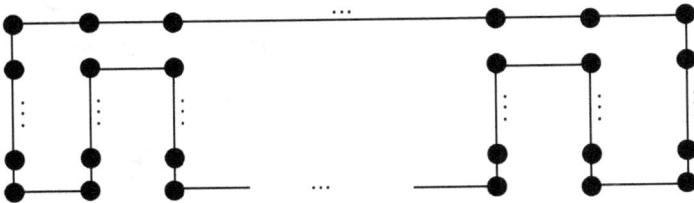

($\Rightarrow$) Suppose that both m and n are odd. Consider G as the graph with

$$V(G) = \{(i,j) \mid i,j \text{ are integers}, 1 \leq i \leq n, 1 \leq j \leq m\}$$

and

$$E(G) = \{uv \mid u = (i,j) \text{ and } v = (i',j') \text{ in } V(G), |i - i'| + |j - j'| = 1\}.$$

Let

$$S = \{(i,j) \in V(G) \mid i + j \text{ is odd}\}$$

as shown below:

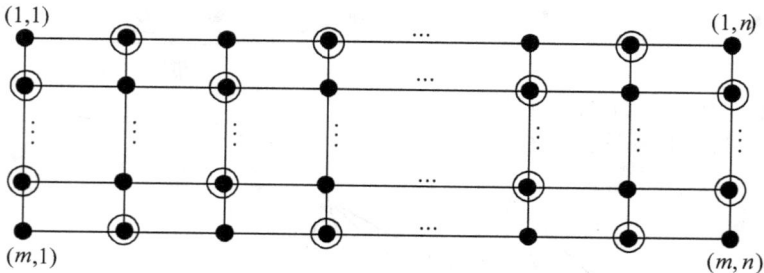

Observe that $|S| = \frac{mn-1}{2}$ and $c(G - S) = \frac{mn+1}{2} = |S| + 1$. Thus, G is not Hamiltonian by Theorem 6.3. $\qquad\square$

Note: *Let G be a bipartite graph with bipartition (X, Y). If G is Hamiltonian, then $|X| = |Y|$ and hence the order of G is even.*

A proof of this fact can be found in the solution of Problem 6 of Exercise 3.1. Another proof by Theorem 6.3 is given below.

Suppose on the contrary that $|X| \neq |Y|$, say $|X| < |Y|$. We then observe that $c(G - X) = |Y| > |X|$. By Theorem 6.3, G is not Hamiltonian, a contradiction.

By the result of this note, we have a simple proof for the necessity of Problem 3.

Let G be the graph for the $m \times n$ rectangular grid. Notice that G is a bipartite graph of order mn. If G is Hamiltonian, then the order of G is even by the above note. Thus mn is even, implying that either m or n is even. $\qquad\square$

Problem 4   Answer: (i) False. (ii) False. (iii) True.

Problem 6   Answer: (i) No. (ii) Yes.

Problem 7   *Solution.*   Let w be the first vertex in traversing the Hamiltonian cycle C of G. We may assume that w is in A. Note that each time we leave A for $V(G)\backslash A$ via an edge in C, as eventually we have to go back to w, there must be another edge in C for us to leave $V(G)\backslash A$ for A. Thus all edges in C linking A and $V(G)\backslash A$ are paired up, and the result follows.

$\square$

Problem 13   *Solution.*   (i) As G is Hamiltonian, $\delta(G) \geq 2$. Let (X, Y) be the bipartition of G. As G is Hamiltonian, $|X| = |Y| = 4$. Thus, $\Delta(G) \leq 4$.

   (ii) Suppose that G contains a Hamiltonian cycle as shown below:

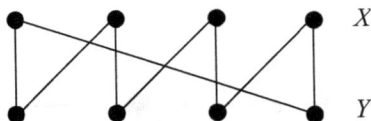

As $\Delta(G) = 4$, there is a vertex of degree 4 as shown below:

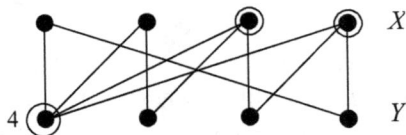

Since G is Eulerian, G contains no odd vertices. It follows that $G \cong K(4, 4)$, as shown in the following sequence of logical implications:

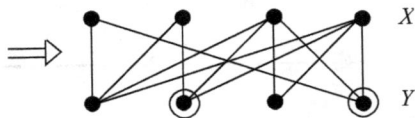

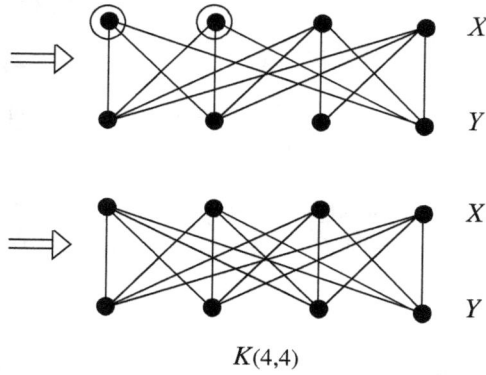

$K(4,4)$

□

Problem 14 *Solution.* (i) No, H is not Hamiltonian. (The reader may apply the method of 'degree two' or Theorem 6.3 to justify it.)
  (ii) $m(H) = 3$.

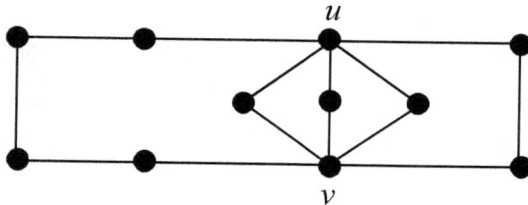

Note that, as shown above, $c(H - \{u, v\}) = 5$. Thus, by adding any two new edges to H to form H^*, we still have $c(H^* - \{u, v\}) \geq 3 > |\{u, v\}|$; that is, H^* is still non-Hamiltonian by Theorem 6.3. It follows that $m(H) \geq 3$. The following example shows that three extra edges are enough.

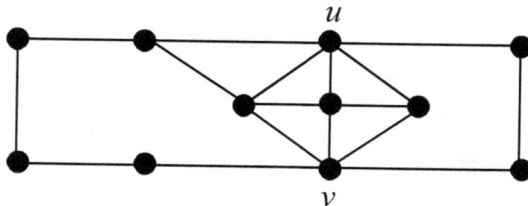

(iii) Another example is given below:

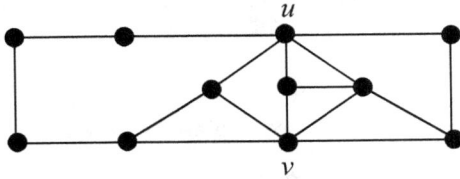

$\square$

Problem 15 *Solution.* (i) No, H is not Hamiltonian. (The reader may apply the method of 'degree two' or Theorem 6.3 to justify it.)

(ii) $m(H) = 2$.

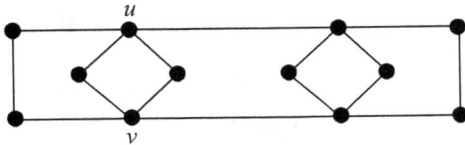

Note that, as shown above, $c(H - \{u, v\}) = 4$. Thus, by adding any one new edge to H to form H^*, we still have $c(H^* - \{u, v\}) \geq 3 > |\{u, v\}|$; that is, H^* is still non-Hamiltonian by Theorem 6.3. It follows that $m(H) \geq 2$. The following example shows that two extra edges are enough.

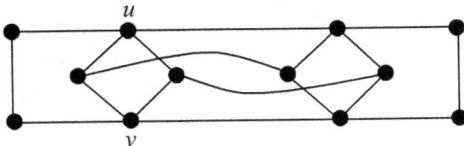

(iii) Another example is given below:

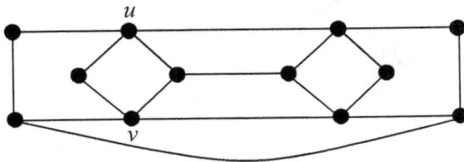

$\square$

Exercise 6.5

Problem 3 *Solution.* (1) (i) is correct, since if (D) holds, then (O) holds.
(2) (ii) is correct.   □

Problem 4 *Solution.* (i) Since u, v are not adjacent,

$$m = e(G - \{u, v\}) + d(u) + d(v) \leq \binom{n-2}{2} + n - 1 = \binom{n-1}{2} + 1.$$

(ii) Since $m \geq \binom{n-1}{2} + 2$, by (i), we have $d(u) + d(v) \geq n$ for every two
non-adjacent vertices u, v in G. Thus, by Theorem 6.5, G is Hamiltonian.
                                                                            □

Problem 6 *Solution.* (i) This graph is not Hamiltonian, but it contains
a Hamiltonian path.

(ii) Let H be the graph obtained from G by adding one new vertex w
and adding n edges joining w to every vertex in G.

Note that the order of H is $n+1$. For every vertex $x \in V(H)$, if $x = w$,
then $d_H(x) = n$; otherwise,

$$d_H(x) = d_G(x) + 1 \geq \frac{n-1}{2} + 1 = \frac{n+1}{2}.$$

By Theorem 6.4, H contains a Hamiltonian cycle C. Observe that $C - w$
is a Hamiltonian path in G.   □

Problem 7 *Solution.* Let $n = 2k + 1$. Let G be the graph obtained
from two K_{k+1}'s by gluing them at one vertex, denoted by w, as shown
below:

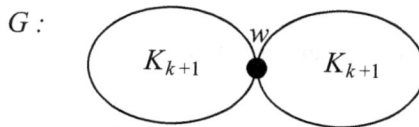

Note that the order of G is $2(k+1) - 1 = 2k+1 = n$, and $d(x) = k = (n-1)/2$
for every vertex x in G, except w.

Since $G - w$ is disconnected, G is not Hamiltonian by Theorem 6.3.   □

Note: $K(k, k+1)$ is another non-Hamiltonian graph G of order n such
that $\delta(G) = (n-1)/2$, where $n = 2k + 1$.

Problem 8 *Solution.* (i) $(3K_2) + N_7$ is not Hamiltonian.

Let S be the set of vertices in $3K_2$. So $|S| = 6$. Observe that if we remove all vertices of S from the graph $(3K_2) + N_7$, we will obtain the graph N_7, which has 7 components. By Theorem 6.3, $(3K_2) + N_7$ is not Hamiltonian.

(ii) $(4K_2) + N_7$ is Hamiltonian. Note that the order of $(4K_2) + N_7$ is 15 and the minimum degree of this graph is 8 ($\geq 15/2$). By Theorem 6.4, $(4K_2) + N_7$ is Hamiltonian. (Indeed this graph is 8-regular.) □

8.7 Selected questions in Chapter 7

Exercise 7.1

Problem 4 *Solution.* (1)

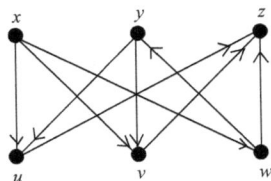

(2) $e(D) = 9$.

(3) z did not win any game.

(4) In team B, w won the largest number of matches. □

Exercise 7.2

Problem 7 *Solution.* Let W be a shortest $u - v$ directed walk in D. We shall show that W is a directed path.

Suppose that W is not a path. Then, W contains a cycle and some vertex in the cycle is repeated. Let x be such a vertex. Let W' be the $u - x$ walk along W in which x is not repeated, and let W'' be the $x - v$ walk along W in which x is not repeated. Then the walk $W_0 = W'W''$ is a $u - v$ walk whose arcs are in W. Note that W_0 is shorter than W, contradicting the assumption of W.

Thus, the conclusion follows. □

Problem 8   *Solution.* Let $P = x_0 x_1 \cdots x_s$ be a longest path in D. As P is a longest path, x_s is not adjacent to any vertex in $V(D) \backslash \{x_0, x_2, \cdots, x_{s-1}\}$.

Thus, as $od(x_s) \geq k$, there exist at least k vertices in $\{x_0, x_1, \cdots, x_{s-1}\}$ which are adjacent from x_s. Hence $s \geq k$, implying that the length of P is at least k.

Similarly, it can be shown that if $k = \min\{id(v) | v \in V(D)\}$, then D contains a path of length at least k.   □

Problem 9   *Solution.* By the result in Problem 8, we know that D contains a longest path $P : x_0 x_1 \cdots x_s$ such that $s \geq k$.

As P is a longest path in D, all vertices adjacent from x_s are in the set $\{x_0, x_1, \cdots, x_{s-1}\}$.

Since $od(x_s) \geq k$, at least k vertices in $\{x_0, x_1, \cdots, x_{s-1}\}$ are adjacent from x_s. Thus there exists a vertex x_i with $0 \leq i \leq s - k$ such that $x_s x_i$ is an arc in D. Hence we get a cycle of length r:

$$x_i x_{i+1} \cdots x_s x_i,$$

where $r = s - i + 1 \geq k + 1$.

Likewise, if $k = \min\{id(v) | v \in V(D)\}$, we can show that the result also holds.   □

Problem 10   *Solution.* If D contains a sink x, i.e., $od(x) = 0$, then no vertex other than x is reachable from x and so D is not strong. Likewise, if D contains a source, then D is not strong.

Hence if D is strong, then $id(v) \geq 1$ and $od(v) \geq 1$ for each vertex v in D.

The converse is not true. The following digraph is an example.

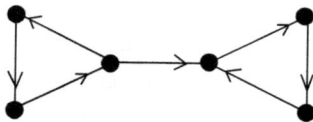

    □

Problem 11 *Solution.* Suppose that D contains no 4-cycles. We first label the vertices in this digraph as shown below:

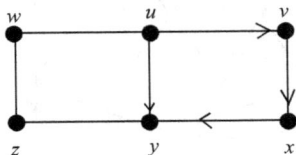

Without loss of generality, assume that $v \to x$. As D contains neither sink nor source, $u \to v$ and $x \to y$. By the assumption that D contains no 4-cycles, $u \to y$, as shown in the diagram.

As D contains no sink, $y \to z$, and in turn, $z \to w$ and $w \to u$. But then $uyzwu$ is a 4-cycle in D, a contradiction.

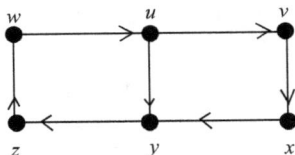

□

Problem 12 *Solution.* Assume that D contains an 8-cycle, but D contains no 4-cycles.

We may assume that $uu'vv'ww'xx'u$ is an 8-cycle as shown below:

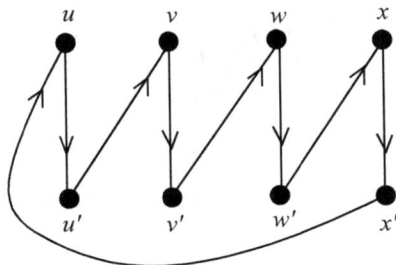

Since D contains no 4-cycles, $u \to v'$; otherwise, $uu'vv'u$ is a 4-cycle. By the same reason, $v \to w'$, $w \to x'$ and $x \to u'$, as shown in the following:

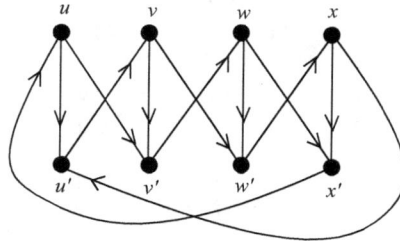

However, we find a 4-cycle, namely, $uv'wx'u$, a contradiction.   □

Problem 13   *Solution.*   We claim that for any two vertices u, v in D, if u is not adjacent to v, then uwv is a path in D for some w in D.

Let $O(u)$ be the set of vertices in D which are adjacent from u and $I(v)$ be the set of vertices in D which are adjacent to v. As u is not adjacent to v, $v \notin O(u)$ and $u \notin I(v)$. So

$$O(u) \cup I(v) \subseteq V(D) \backslash \{u, v\}.$$

If $O(u) \cap I(v) = \emptyset$, then

$$od(u) + id(v) = |O(u)| + |I(v)| \leq |V(D) \backslash \{u, v\}| = n - 2,$$

a contradiction. Thus there exists $w \in O(u) \cap I(v)$, implying that uwv is a path in D.   □

Problem 14   *Solution.*   Suppose on the contrary that D contains no sink; i.e., $od(v) > 0$ for every vertex v in D.

Let $k = \min\{od(v) | v \in V(D)\}$. Then $k \geq 1$. By the result in Problem 9, D contains an r-cycle, where $r \geq k + 1$, a contradiction.

Similarly, if $id(v) > 0$ for every vertex v in D, then D also contains a cycle.

Hence the result holds.

The converse is not true. The following digraph contains a source and a sink, and also a cycle.

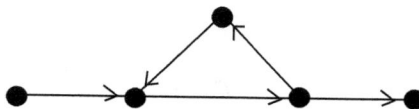

□

Problem 15   *Solution.*   ($\Leftarrow$) If D contains a cycle, say $v_1 v_2 \cdots v_k v_1$, then $v_1 v_2 \cdots v_k v_1 v_2$ is a $v_1 - v_2$ walk which is not a path.

($\Rightarrow$) Let W be a walk in D that is not a path. Then some vertex is repeated in W. Let v be a vertex in W which is repeated. We may choose v so that no vertex in the section (*) (see the diagram below) is repeated in (*).

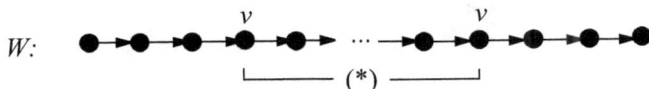

Clearly, the closed walk of (*) forms a cycle in D.   □

Problem 16   *Solution.*   The following digraph is such a digraph.

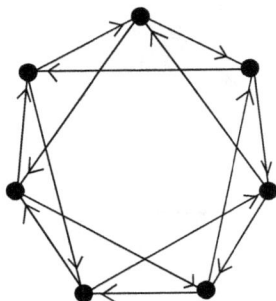

Problem 17   *Solution.*
(a) (i) $R(S) = \{x, z, y\}$, $R(T) = \{u, v, w, x, y, z\}$, $R(U) = \{x, y, z\}$.
   (ii) Yes, $R(U) = U$.
   (iii) D is not strong.

(b) (i) No, there is no non-empty subset W of vertices in D such that $W \neq V(D)$ and $R(W) = W$.
   (ii) Yes, D is strong.

(c) ($\Rightarrow$) Assume that D is strong. Then any two vertices are mutually reachable in D. Thus for any W, $\emptyset \neq W \subset V(D)$, we have $R(W) = V(D)$. Hence the necessity holds.

($\Leftarrow$) Suppose that D is not strong. Then there exist two vertices u, v in D such that u is not reachable from v.

Let $W = R(\{v\})$. Then W is a non-empty and proper subset of $V(D)$.

We claim that $R(W) = W$. Let $x \in R(W)$. Then x is reachable from a vertex y in W. As $y \in W = R(\{v\})$, y is reachable from v. Hence x is reachable from v, i.e., $x \in R(\{v\}) = W$. This shows that $W = R(W)$.

So the sufficiency holds.   □

Problem 18   *Solution.*   (i) For any graph G of order at least 2, there is no orientation D such that $d(x, y) \leq 1$ for all x, y in $V(D)$.

The reason is very simple. If $d(x, y) = 1$, then it is impossible that $d(y, x) = 1$.

(ii) An orientation D_n of K_n, $n = 3, 5, 6$, such that $d(x, y) \leq 2$ for any $x, y \in V(D_n)$, is given below:

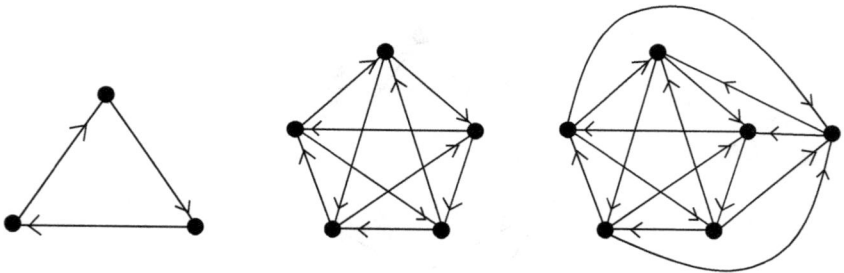

(iii) An orientation D_4 of K_4 such that $d(x, y) \leq 3$ for any $x, y \in V(D_4)$ is given below:

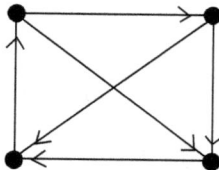

(iv) No, there is no orientation D_4 of K_4 such that $d(x, y) \leq 2$ for any $x, y \in V(D_4)$.

Suppose on the contrary that such a D_4 exists. Let yx be an arc in D_4.

Then $d(x, y) = 2$, and so $xzyx$ is a 3-cycle in D_4 for some vertex z as shown in (a).

Let u be the fourth vertex. Then u must be adjacent to some vertex in $\{x, y, z\}$, say x (by symmetry). Then $d(x, u) = 2$, implying that $z \to u$, as shown in (b).

(a)   (b)

Now consider y and u. If $y \to u$, then $d(u, y) = 3$; if $u \to y$, then $d(y, u) = 3$, a contradiction.

(v) An orientation D_{n+2} of K_{n+2} is constructed from D_n by adding two new vertices u and v and adding all arcs in the set

$$\{uv\} \cup \{vw, wu | w \in V(D_n)\},$$

as shown in the following.

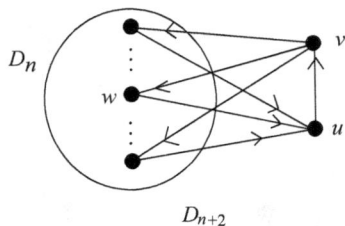

D_{n+2}

Now we show that $d(x, y) \leq 2$ for all $x, y \in V(D_{n+2})$. This is obvious if $\{x, y\} \cap \{u, v\} = \emptyset$.

By the definition of D_{n+2}, $uvwu$ is a 3-cycle of D_{n+2} for every $w \in V(D_n)$. Thus $d(u, v) = 1$, $d(v, u) = 2$ and

$$d(u, w) = 2, d(w, u) = 1, d(v, w) = 1, d(w, v) = 2,$$

for every $w \in V(D_n)$. Hence $d(x, y) \leq 2$ for all $x, y \in V(D_{n+2})$ if $\{x, y\} \cap \{u, v\} \neq \emptyset$. □

Problem 19   *Solution.*   (i) As D is strongly connected and D contains only 5 vertices, $d(x, y) \leq 4$ for all x, y in $V(D)$.

(ii) No, we have $d(a, c) = 4$ (see the following diagram).

D:
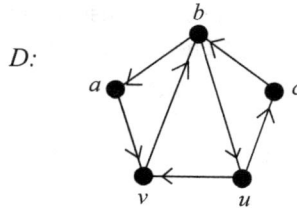

(iii) The following diagram shows an orientation D' such that $d(x, y) \leq 3$ for all x, y in $V(D')$.

D':
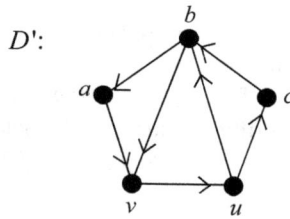

(iv) No, the graph G has no orientation D such that $d(x, y) \leq 2$ for all x, y in $V(D)$.

G:
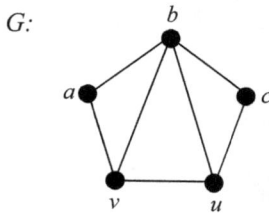

(a)

D:
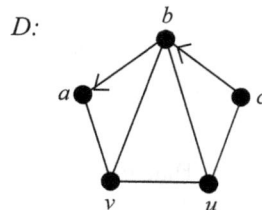

(b)

Suppose on the contrary that such a D exists. Consider a and c. As $d(c, a) \leq 2$, we must have $c \rightarrow b \rightarrow a$ (see the above diagram) in D. But then $d(a, c) \leq 2$ can never be the case, a contradiction.   □

Problem 20   *Solution.*   (i) There is no orientation D of G such that $d(x, y) \leq 2$ for all x, y in $V(D)$. Let x, y be two vertices in D with $x \to y$. Then $d(y, x) \geq 2$. If $d(y, x) = 2$, then there exists a 3-cycle $xywx$ in D, contradicting the fact that G is bipartite.

   (ii) Let (X, Y) be the bipartition of $K(p, p)$, where $X = \{a_1, a_2, \cdots, a_p\}$ and $Y = \{b_1, b_2, \cdots, b_p\}$. Let D be the orientation with arc set:

$$\{a_i b_i \mid i = 1, 2, \cdots, p\} \cup \{b_i a_j \mid 1 \leq i, j \leq p, i \neq j\}.$$

For any i, j with $i \neq j$, $a_i b_i a_j b_j a_i$ is a 4-cycle. Thus $d(a_i, a_j) \leq 3$, $d(b_i, b_j) \leq 3$, $d(a_i, b_j) \leq 3$ and $d(b_i, a_j) \leq 3$. Hence $d(x, y) \leq 3$ in D for all x, y in $V(D)$.   □

Problem 21   *Solution.*   (i) The digraph is shown below:

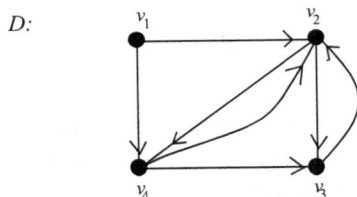

D:

(ii) The adjacency matrix of D is $A(D) = \begin{pmatrix} 0 & 1 & 1 & 0 & 1 \\ 0 & 0 & 1 & 1 & 0 \\ 0 & 0 & 0 & 1 & 0 \\ 0 & 0 & 0 & 0 & 1 \\ 1 & 0 & 0 & 1 & 0 \end{pmatrix}$.   □

Problem 22   *Solution.*   (i) If $od(v) \geq 1$ for all $v \in V(D)$, then, by the result of Problem 9, D contains a cycle of length at least 2, a contradiction.

   (ii) ($\Rightarrow$) Assume that D contains no cycles.
   We shall prove the result by induction on $n \geq 2$. For $n = 2$, D is either

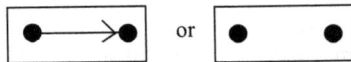

or

Name the vertices of D as

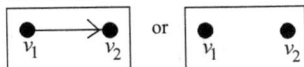

or

Then

$$A(D) = \begin{pmatrix} 0 & 1 \\ 0 & 0 \end{pmatrix} \text{ or } A(D) = \begin{pmatrix} 0 & 0 \\ 0 & 0 \end{pmatrix}.$$

Now assume that $n \geq 3$. By the result in (i), there is a vertex v such that $od(v) = 0$. Label this vertex as v_n.

Let D' be the digraph obtained from D by deleting v_n. It is clear that D' contains no cycles. Thus, by the induction hypothesis, the vertices in D' can be named as $v_1, v_2, \cdots, v_{n-1}$ such that $A(D')$ is upper triangular. Hence $A(D)$ is also upper triangular, as

$$A(D) = \left(\begin{array}{c|c} A(D') & \begin{array}{c} a_{1,n} \\ \vdots \\ a_{n-1,n} \end{array} \\ \hline 0 \cdots 0 & 0 \end{array} \right).$$

($\Leftarrow$) Assume that the vertices in D can be named as $v_1, v_2, \cdots, v_n$ such that $A(D)$ is upper triangular. Then the (i,j)-entry in $A(D)$ is 0 if $i \geq j$, implying that $v_i v_j$ is not an arc in D whenever $i \geq j$. If D contains a cycle $v_{i_1} v_{i_2} \cdots v_{i_k} v_{i_1}$, then $i_1 < i_2 < \cdots < i_k < i_1$, which is impossible. $\qquad \square$

Exercise 7.3

Problem 2   *Solution.*   As $d(u,v) = k \geq 2$, let $u_1 \cdots u_k v$ be a shortest $u - v$ path in T, where $u_1 = u$. Then $v \to u_i$ for each $i = 1, 2, \cdots, k-1$. Thus, we have

(i) $od(v) \geq k - 1$;

(ii) for $r = 3, 4, \cdots, k+1$, $v u_{k-r+2} u_{k-r+3} \cdots u_k v$ is a r-cycle that contains v;

(iii) $v u_1 u_2 \cdots u_k v$ is a $(k+1)$-cycle that contains both u and v. $\qquad \square$

Problem 9   *Solution.*   Let T be of order n. By Result (2), we have $od(v) + id(v) = n - 1$ for each vertex v in T, and

$$\sum_{v \in V(T)} od(v) = \binom{n}{2} = \sum_{v \in V(T)} id(v).$$

Thus

$$\sum_{v \in V(T)} (od(v))^2 - \sum_{v \in V(T)} (id(v))^2 = \sum_{v \in V(T)} ((od(v))^2 - (id(v))^2)$$

$$= \sum_{v \in V(T)} (od(v) - id(v))(od(v) + id(v))$$

$$= (n-1) \sum_{v \in V(T)} (od(v) - id(v))$$

$$= (n-1) \left(\sum_{v \in V(T)} od(v) - \sum_{v \in V(T)} od(v) \right)$$

$$= (n-1) \left(\binom{n}{2} - \binom{n}{2} \right)$$

$$= 0.$$

The result holds. $\square$

Problem 10   *Solution.*   Let T_k be the sub-tournament of T induced by $\{v_1, v_2, \cdots, v_k\}$. Then, for $i = 1, 2, \cdots, k$,

$$od(v_i) \geq od_{T_k}(v_i),$$

where $od_{T_k}(v_i)$ is the out-degree v_i in T_k. Hence

$$\sum_{i=1}^{k} od(v_i) \geq \sum_{i=1}^{k} od_{T_k}(v_i) = \binom{k}{2}.$$ $\square$

Problem 13   *Solution.*   ($\Rightarrow$) Suppose that T does not contain two vertices of the same out-degree. Then the out-degrees of vertices in T are

$$0, 1, 2, \cdots, n-1.$$

Let $V(T) = \{x_1, x_2, \cdots, x_n\}$ and $od(x_i) = i - 1$. It can be shown by induction that

$$E(T) = \{x_j x_i \mid 1 \leq i < j \leq n\}.$$

Thus T contains no cycles, a contradiction.

($\Leftarrow$) Assume that T contains two vertices of the same out-degree.

Let x and y be two vertices in T with $od(x) = od(y) = k$. As either $x \to y$ or $y \to x$, $k \geq 1$. Assume that $x \to y$ and y is adjacent to k vertices $y_1, y_2, \cdots, y_k$. Since $od(x) = od(y) = k$ and $x \to y$, it is impossible that x

is adjacent to every vertex in $\{y_1, y_2, \cdots, y_k\}$. Assume that $x \nrightarrow y_1$. Then $y_1 \rightarrow x$, and so xyy_1x is a 3-cycle in T. ☐

Problem 14   *Solution.* ($\Rightarrow$) Assume that T is transitive, i.e., for any three vertices u, v, w, if $u \rightarrow v$ and $v \rightarrow w$, then $u \rightarrow w$. Thus T contains no 3-cycles. By the result of Problem 13, $od(u) \neq od(v)$ for any two vertices u, v in T.

($\Leftarrow$) Assume that $od(u) \neq od(v)$ for any two vertices u, v in T. By the result of Problem 13, T contains no 3-cycles. Thus, for any three vertices u, v, w in T, if $u \rightarrow v$ and $v \rightarrow w$, we must have $u \rightarrow w$; that is, T is transitive. ☐

Problem 15   *Solution.* By the result of Problem 14, T is transitive if and only if $od(u) \neq od(v)$ for every two vertices u, v in T. Since T is of order n and $0 \leq od(u) \leq n - 1$ for all $u \in V(T)$, $od(u) \neq od(v)$ for every two vertices u, v in T if and only if the out-degrees of vertices in T are, respectively, $n - 1, n - 2, \cdots, 1, 0$. ☐

Problem 16   *Solution.* Notice that if T has a vertex of in-degree 0 or out-degree 0, then T is reducible.

(i) This tournament T is reducible, as it contains a vertex of in-degree 0.

(ii) Let T be a transitive tournament of order n. By the result of Problem 15, the out-degrees of vertices in T are $n - 1, n - 2, \cdots, 1, 0$. As T contains a vertex of out-degree 0, T is reducible. ☐

Exercise 7.4

Problem 1   *Solution.* ($\Rightarrow$) Let T be a tournament of order n. The proof is by induction on n. Assume that T is transitive. If $n \leq 2$, then it is obvious that T contains one and only one Hamiltonian path.

Now assume that $n \geq 3$.

Since T is transitive, by the result of Problem 15 in Exercise 7.3, T has a vertex v with $od(v) = n - 1$. As $T - v$ is also a transitive tournament, by the induction hypothesis, $T - v$ contains one and only one Hamiltonian path. Since v is adjacent to every vertex in $T - v$, T also contains one and only one Hamiltonian path.

($\Leftarrow$) Let $x_1 x_2 \cdots x_n$ be the unique Hamiltonian path in T. We shall show that $x_i \to x_j$ for all i, j with $1 \leq i < j \leq n$.

Clearly, it holds if $n \leq 2$.

Assume that $x_i \to x_j$ for all i, j with $1 \leq i < j \leq n - 1$. We need to show that $x_i \to x_n$ for all i with $1 \leq i \leq n - 1$.

Suppose on the contrary that $x_n \to x_i$ for some i with $1 \leq i \leq n - 1$. Let k be the minimum integer with $1 \leq k \leq n - 1$ such that $x_n \to x_k$.

If $k = 1$, then $x_n x_1 x_2 \cdots x_{n-1}$ is also a Hamiltonian path in T, a contradiction.

If $k \geq 2$, then $x_1 x_2 \cdots x_{k-1} x_n x_k x_{k+1} \cdots x_{n-1}$ is a Hamiltonian path in T, also a contradiction.

Hence $x_i \to x_n$ for all i with $1 \leq i \leq n - 1$.

Therefore $x_i \to x_j$ for all i, j with $1 \leq i < j \leq n$. This shows that T is transitive. $\qquad \square$

Problem 2 *Solution.* We claim that if $od(u) \geq od(v)$, then $d(u, v) \leq 2$.

If $u \to v$, then $d(u, v) = 1$. Assume that $v \to u$. We shall show that $d(u, v) = 2$.

Let $od(u) = k$ and assume that $u \to u_i$ for $i = 1, 2, \cdots, k$. As $od(v) \leq k$ and $v \to u$, it is impossible that $v \to u_i$ for all $i = 1, 2, \cdots, k$. Thus $u_i \to v$ for some i, where $i = 1, 2, \cdots, k$, say $u_1 \to v$. Then $u u_1 v$ is a $u - v$ path, implying that $d(u, v) = 2$ in this case. $\qquad \square$

Problem 3 *Solution.* If there are even number of teams in the round-robin tournament, only (i) is possible; otherwise, both (i) and (ii) are possible. But (iii) is impossible.

Assume that there are n teams in the round-robin tournament. Then $p + q = n - 1$ and
$$p \geq \binom{n}{2} / n = \frac{n-1}{2}.$$

Thus $q \leq (n-1)/2$. Hence $p \geq q$.

If n is even, we have $p \geq \lceil (n-1)/2 \rceil \geq n/2 > q$. If n is odd, then it is possible that $p = q = (n-1)/2$ or $p > (n-1)/2 \geq q$. $\qquad \square$

Problem 5 *Solution.* Assume that $od(u) = k$, where $0 \leq k \leq n - 2$.

Let $V(T) = \{u, x_1, x_2, \cdots, x_{n-1}\}$. Assume that $u \to x_i$ for $i = 1, 2, \cdots, k$, and $x_i \to u$ for $i = k+1, k+2, \cdots, n-1$.

Let T' be the sub-tournament of T induced by $\{x_{k+1}, x_{k+2}, \cdots, x_{n-1}\}$. Then T' itself has a king, say x_{n-1}.

So $d(x_{n-1}, x_i) \leq 2$ for all $i = k+1, k+2, \cdots, n-2$. Note also that $x_{n-1} \to u \to x_i$ for all $i = 1, 2, \cdots, k$. Thus x_{n-1} is a king of T.

This shows that u is dominated by a king of T.   □

Problem 7   *Solution.*   No, no tournament contains exactly two kings.

Suppose on the contrary that T is a tournament of order $n \geq 2$ which contains exactly two kings, say u and v with $u \to v$. Since u is reachable from v, $od(u) \leq n-2$ and $n \geq 3$. By the result of Problem 5, u is dominated by a king, say w, in T. Clearly, $w \neq v$. Thus, T has at least three kings, namely, u, v and w, a contradiction.   □

Problem 12   *Solution.*   (i) This tournament is irreducible.

(ii) ($\Rightarrow$) Assume that T is a strong tournament. Let (X, Y) be any partition of $V(T)$.

Let $x \in X$ and $y \in Y$. As T is strong, there exists a $x - y$ path in T. This path must contain an arc $x'y'$ with $y' \in Y$ and $x' \in X$.

Similarly, there exists an arc $y''x''$ with $y'' \in Y$ and $x'' \in X$.

Hence T is irreducible.

($\Leftarrow$) Assume that T is irreducible. To show that T is strong, we show that $R(x) = V(T)$ for all vertices x in T, where $R(x)$ is the set of all vertices in T which are reachable from x.

Suppose that $R(x)$ is a proper subset of $V(T)$ for some $x \in V(T)$. Since T is irreducible, there exists an arc $x'y'$ in T with $x' \in R(x)$ and $y' \in V(T) \backslash R(x)$. But this implies that $y' \in R(x)$, a contradiction. Hence $R(x) = V(T)$, and so T is a strong tournament.   □

Problem 14   *Solution.*   (i) It is true. Let z be any vertex in T. Let

$$O(z) = \{x \in V(T) \mid z \to x\} \text{ and } I(z) = \{y \in V(T) \mid y \to z\}.$$

Since T is strong, both $O(z)$ and $I(z)$ are non-empty. Also, there exist $x \in O(z)$ and $y \in I(z)$ such that $x \to y$; otherwise, there is no path from z to any vertex in $I(z)$. Clearly, $zxyz$ is a 3-cycle containing z, as shown below.

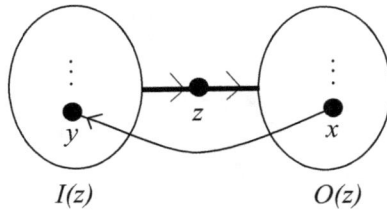

I(z) O(z)

(ii) It is false.

The following tournament T of order 4 is a strong tournament.

Note that the arc xy is not contained in a 3-cycle. So (ii) is false.

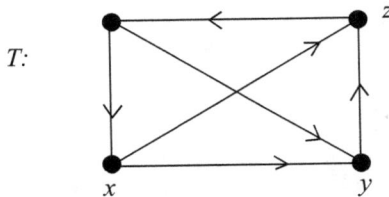

(iii) It is false.

In the tournament T shown in (ii), the arc xz is not contained in a Hamiltonian cycle.

(iv) It is true.

Let xy be any arc in T. As T is strong, there exists a $y - x$ path P in T. Clearly, P and xy form a cycle which contains xy.

(v) It is false.

In the tournament T shown in (ii), there is no Hamiltonian path from x to z, nor Hamiltonian path from z to x. □

References

[B] J A Bondy, A graph reconstructor's manual, in: Surveys in Combinatorics, Proceedings of the 13th British Combinatorial Conference, Guildford, UK, 1991, London Mathematical Society, Lecture Note Series, **166**(1991), 221 – 252.

[C] P Camion, Chemins et circuits hamiltoniens des graphes complets, *C R Acad. Sciences, Paris*, **249**(1959) 2151 – 2152.

[E] L Euler, The solution of a problem relating to the geometry of position, *Commentarii Academiae Scientiarum Imperialis Petropolitanae* **8**(1736), 128 – 140.

[F] S Fortin, The graph isomorphism problem, Technical Report TR 96-20, University of Alberta, 1996.

[H] C Hierholzer, On the possibility of traversing a line-system without repetition on discontinuity, *Mathematische Annalen* **6**(1873), 30 – 32.

[Ke] P J Kelley, On isometric transformations, Ph.D. Thesis, University of Wisconsin, 1942.

[KST] J Kobler, U Schoning and J Toran, The Graph Isomorphism Problem — Its Structural Complexity, Springer Verlag, Birkhäuser, 1993.

[Kö] D König, Theorie der endlichen und unendlichen Graphen, Akademische Verlags-gesellschaft, Leipzig, 1936.

[L] H G Landau, On dominance relations and the structure of animal societies III, the condition for a score sequence, *Bull. Math. Biophys.*, **15**(1955), 143 – 148.

[LO] L. Lovász, Three short proofs in graph theory, *Journal of Combinatorial Theory, Series B* **19** (3) (1975), 269 – 271, doi:10.1016/0095-8956(75)90089-1.

[M] S B Maurer, The king chicken theorems, *Math. Magazine*, **53**(1980), 67 – 80.

[P] J Petersen, On the theorem of Tait, *L'Intermédiaire des Mathématiciens* **5** (1898), 225 – 227.

[R] L Rédei, Ein Kombinatorischer Satz, *Acta Litt. Szeged* **7**(1934), 39 – 43.

[RSST] N Robertson, D P Sanders, P D Seymour, and R Thomas, The Four Colour Theorem, *J. Comb. Theory, Series B*, **70**(1997), 2 – 44.

[U] S M Ulam, A collection of mathematical problems, Wiley Interscience, New York, 1960 (p. 29).

[W] R Wilson, Four colors suffice: How the map problem was solved, Princeton University Press, 2004.

Books Recommended

(1) J. M. Aldous and R. J. Wilson, *Graphs and Applications: An Introductory Approach*, New York: Springer, 2000.

(2) F. Buckley and M. Lewinter, *A Friendly Introduction to Graph Theory*, Prentice Hall, 2003.

(3) G. Chartrand, *Graphs as Mathematical Models*, Prindle, Weber & Schmidt, 1977.

(4) G. Chartrand and L. Lesniak, *Graphs and Digraphs*, Chapman and Hall, 1996.

(5) G. Chartrand and O. R. Oellermann, *Applied and Algorithmic Graph Theory*, New York: McGraw-Hill, 1993.

(6) G. Chartrand and P. Zhang, *Introduction to Graph Theory*, New York: McGraw-Hill, 2004.

(7) J. Clark and D. A. Holton, *A First Look at Graph Theory*, World Scientific, 1991.

(8) J. Gross and J. Yellen, *Graph Theory and Its Applications*, CRC Press, 1999.

(9) F. Roberts and B. Tesman, *Applied Combinatorics*, 2nd edition, Pearson/Prentice Hall, 2005.

(10) P. Tannenbaum and R. Arnold, *Excursions in Modern Mathematics*, 5th edition, Prentice Hall, 2004.

(11) W. D. Wallis, *A Beginner's Guide to Graph Theory*, Birkhäuser, 2006.

(12) R. J. Wilson and J. J. Watkins, *Graphs: An Introductory Approach*, New York: Wiley, 1990.

Index